INTRODUCTION TO TELECOMMUNICATIONS NETWORKS

INTRODUCTION TO TELECOMMUNICATIONS NETWORKS

Gordon F. Snyder Jr.

National Center for Telecommunications Technologies
Springfield Technical Community College

THOMSON

DELMAR LEARNING

Australia Canada Mexico Singapore Spain United Kingdom United States

THOMSON

DELMAR LEARNING

Introduction to Telecommunications Networks

Gordon F. Snyder Jr.

Executive Director:
Alar Elken

Executive Editor:
Sandy Clark

Senior Acquisitions Editor:
Gregory L. Clayton

Senior Developmental Editor:
Michelle Ruelos Cannistraci

Executive Marketing Manager:
Maura Theriault

Channel Manager:
Fair Huntoon

Marketing Coordinator:
Brian McGrath

Executive Production Manager:
Mary Ellen Black

Production Manager:
Larry Main

Senior Project Editor:
Christopher Chien

Art/Design Coordinator:
David Arsenault

Editorial Assistant:
Dawn Daugherty

**STCC Foundation Press
Executive Director:**
Debbie Bellucci

**Library of Congress
Cataloging-in-Publication Data**

Snyder, Gordon F., Jr.
   Introduction to telecommunications networks / Gordon F. Snyder Jr.
      p. cm.
   ISBN 1-4018-6486-4
   1. Telecommunication systems.
I. Title.
   TK5101.S59 2002
   621.382'1—dc21   2002075054

NOTICE TO THE READER

Dedicated to my dad,
Gordon F. Snyder Sr.

Contents

Preface

Audience

This book is designed as an introductory text for telecommunications and information technology students interested in learning about the telecommunications field. The text has been written at a basic introductory level, providing a good overview of telecommunications networks without delving too deeply into any one topic. Prerequisites are minimal, and an algebra background will suffice. It has been designed for the freshman-level Electronics Engineering, Electronics Technology, Computer Information Systems, Telecommunications, and Computer Science student who has not taken advanced mathematics or electronics courses. Hopefully, by completing the text, students' interest in telecommunications will be sparked; they will want to learn more about telecommunications; and they will take more advanced, higher-level courses.

Approach

The text gives basic coverage of the existing public switched telephone network (PSTN), along with the cable modem system, digital subscriber line, and data network systems. Topics are introduced at a basic level and covered without a lot of heavy mathematics. The text provides practical examples of theory, which is presented using mathematics only when required; students will finish with a good understanding of how the existing network has evolved and how it will continue to change as higher-bandwidth network requirements emerge.

Organization

The public switched telephone network (PSTN) has, over the years, continued to change and integrate itself deeper and deeper into our lives. The text covers the emergence and growth of the PSTN and encourages students to consider the future. Students should leave with a good understanding of how and why things have changed in the past, in preparation for changes that will continue into the future. A chapter-by-chapter summary of the text follows:

- *Chapter 1* reviews a brief history of the telecommunications industry, listing the major historical points in the development of the network.

- *Chapter 2* covers the basic telephone set and its four basic functions: to provide a signal that a call has started or a call is finished, to provide the number being called to the connected telephone switch, to indicate there is an incoming call, and to convert voice to electrical frequencies and electrical frequencies to voice. Telephone network connection is also covered, including an introduction to the central office (CO) configuration.

- *Chapter 3* continues to define the frequency range to which the telephone local loop has been tuned; provides more detail on transmission problems on the local loop, including bridged taps; and continues to show how the local loop has been tuned to a selected frequency range using loading coils. This chapter also moves the student beyond the local loop and describes how analog voice signals can be converted to digital. Analog and digital multiplexing are covered, and the digital time division and statistical time division methods are stressed. Wavelength division multiplexing is also explained. The digital signal (DS) level system is covered, and the North American T-carrier and European E-carrier systems are explained and contrasted. The DS level system is further reviewed describing some of the issues involved in transmitting data over a network designed for voice. Chapter 3 concludes with coverage of the synchronous optical network (SONET) system, frame relay, and asynchronous transfer mode (ATM).

- *Chapter 4* covers the various types of telecommunications transmission media, including copper wire, coaxial cable, wireless, and fiber-optic systems. Different transmission media advantages and disadvantages are contrasted. The wire color code is reviewed, transmission systems are modeled in block diagram format, and crosstalk is covered.

- *Chapter 5* covers switching on the PSTN. Switching is an automated process today, but to really understand how it works requires the student to go back and review early techniques. The early human operator switched network is briefly discussed, along with the evolution of mechanical and then computerized switches. Students also get a better understanding of the differences between local and long-distance telephone service. Telephone switch traffic is discussed, and students learn to do some basic call traffic and call blocking calculations.

- *Chapter 6* covers signaling on the PSTN. Signaling, on the voice network, can be considered as anything that is not voice that is passed through the network. For example, picking up a telephone receiver generates a current on the local loop and signals the attached CO switch that a line has been picked up. When making long-distance calls, in addition to local loop signaling, long-distance signaling is required between the caller's local switch and the local switch of the person the caller is trying to call. The chapter continues to describe in more detail how calls move from one point to another and how both in-band signaling and out-of-band signaling are now used on the PSTN. The out-of-band signaling process is described in detail, along with the current Signaling System 7 (SS7) out-of-band signaling method used today. The chapter finishes with a discussion of integrated services digital network (ISDN) and how the ISDN integrates into the SS7 network.

- *Chapter 7* looks at how digital data is modulated and demodulated for transmission across the existing voice designed network. Discussion includes how an analog modem works and identifies the three primary functions of a modem. Four different types of modulation are described: amplitude-shift keying (ASK), frequency-shift keying (FSK), phase-shift keying (PSK), and quadrature amplitude modulation (QAM). Different analog modem modulation standards are covered, including nonstandardized defacto standards. Noise is defined, and calculations and reporting methods are defined for a transmission system. Additional calculations are demonstrated, including the maximum data bit rate given a transmission system bandwidth and also signal-to-noise ratio. Modem error detection and correction are covered, along with different data-compression techniques.

- *Chapter 8* investigates how the Internet bandwidth demand continues to grow and how the existing voice bandwidth local loop is not sufficient. The telephone companies are now using digital subscriber line (DSL) to extend the life of the local loop copper wire pairs. In addition, the cable television companies have discovered that, with some infrastructure changes, they can provide high-bandwidth broadband data and voice along with cable television on the same transmission media. In this chapter, students further understand the bandwidth limitations of the existing local loop and learn why loading coils and bridged taps must be removed to increase the bandwidth of the local loop. Students learn about asymmetric DSL (ADSL) and why it is called *asymmetric*. They also learn the specific frequencies used in ADSL systems. Also, DSL modem function is covered and connection to the Internet at the CO is presented. ADSL signal modulation is covered and other forms of

DSL reviewed, followed by a review of the cable television system. Frequency allocation, bandwidth sharing, modulation methods, privacy, asymmetry, and how the cable system infrastructure has been modified to provide two-way data to neighborhoods are all covered. Peer-to-peer networks are covered, and the concept of bottlenecked data traffic is introduced. Firewalls and proxy servers are discussed, and examples are given. Finally, cable network and DSL systems are contrasted.

- *Chapter 9* starts with an overview of data networks, including local, campus, metropolitan, and wide-area networks. An explanation of peer-to-peer and server-based networks is provided, giving information about where each type would be used. Different network physical topologies are then covered with a brief introduction of each one. Finally, the chapter finishes with a brief introduction to the OSI model and how each layer interacts with layers on local and remote communicating devices.

- *Chapter 10* covers some of the more common layer-2, local-area-network protocols, including Ethernet and Token Ring. Ethernet is covered first with a good review of the 802.3 specification. The chapter then moves into 10Base5 and 10Base2 local-area networks and moves through the evolution of Ethernet to the gigabit-per-second standard known as Gigabit Ethernet. The 568 termination standards are covered, along with crossover cabling and uplinking. Token Ring is covered next with an overview of how the network works and then coverage of how a Token Ring local-area network is extended. The chapter finishes with a discussion of fiber-distributed-data interface (FDDI).

- Included as an *appendix* is a brief review of electronic filter theory and the effects of inductance and capacitance in different locations in filter circuits. This is extremely important to the student, because the PSTN local loop (last mile, final mile, etc.) copper wire pair we all have coming into our homes and businesses for voice telephone service is tuned to be a low-pass filter. By tuning the local loop this way, lower voice frequencies are passed through the network with very high quality.

Completion of this introductory telecommunications course—with this text and its corresponding lab manual—prepares students for more advanced telecommunications courses.

Features

- Information is presented from the technician's point of view, allowing readers to make immediate connections between theory and practice.
- Examples and illustrations focus exclusively on telecommunications concepts without straying into electronics communications topics.

- Objectives, outlines, key-term lists, summaries, and review and discussion questions in every chapter focus attention on key concepts and speed learning.

- An end-of-text glossary includes acronyms and gives students a quick reference to basic terms used in the telecommunications industry.

- Math is limited to algebra and basic trigonometry functions, making information accessible to technology students and to readers without higher-level math skills.

- A companion Lab Manual contains modern telecommunications experiments that may be performed with inexpensive equipment to enhance learning.

Supplements

Instructor's guide. Gives end-of-chapter solutions and answers.
ISBN: 1401862888.

Laboratory manual. Gives a complete semester-long set of experiments that can be performed; chapter lecture material is covered.
ISBN: 1401859244

Web Tutor. This student study guide and interactive supplement offers notes, flashcards, Web links, quizzes, and discussion-room topics. Web Tutor is available using:

Web Tutor on WebCT   ISBN: 1401884105
Web Tutor on Blackboard   ISBN: 1401832334

On-line companion. See the *Telecommunications Infrastructure* on-line companion at http://www.electronictech.com for up-to-date information on evolving telecommunications technologies, along with an updated list of the Web links used throughout the text companion.

National Center for Telecommunications Technologies

The National Center for Telecommunications Technologies (NCTT: http://www.nctt.org) is a National Science Foundation (NSF: http://www.nsf.gov) sponsored Advanced Technological Education (ATE) Center established in 1997 by Springfield Technical Community College (STCC: http://www.stcc.edu) and the National Science Foundation (NSF). All material produced as part of the NCTT textbook series is based on work supported by the Springfield Technical Community College and the National Science Foundation under Grant Number DUE 9751990.

The NCTT was established in response to the telecommunications industry and the worldwide demand for instantly accessible information. Voice, data, and video communications across a worldwide network are creating opportunities that did not exist a decade ago; preparing a workforce to compete in this global marketplace is a major challenge for the telecommunications industry. As we enter the 21st century, with even more rapid breakthroughs in technology anticipated, education is the key and the NCTT is working to provide the educational tools employers, faculty, and students need to keep the United States competitive in this evolving industry.

We encourage you to visit the NCTT Web site, along with the NSF Web site, to learn more about this and other exciting projects. Together we can explore ways to better prepare quality technological instruction and ensure the globally competitive advantage of telecommunications industries in the United States.

Acknowledgments

The author would like to thank the following individuals whose vision and insight helped to provide the motivation for development of this text: From the National Science Foundation: Program Officers Elizabeth Teles and Gerhard Salinger. From Springfield Technical Community College: President Andrew Scibelli, retired Vice President John Dunn, Professor Diane D. Snyder, and Foundation Press Director Debbie Bellucci. From the National Center for Telecommunications Technologies at Springfield Technical Community College: Gary Mullett, James Downing, Nina Laurie, Joseph Joyce, Helen Wetmore, Scott SaintOnge, Fran Smolkowicz, Sandy Rheaume, and retired Director James Masi. From Delmar: Michelle Ruelos Cannistraci and Greg Clayton. From Shepherd Incorporated: Larry Goldberg and his team. Without the help and support of each, this text would never have been completed.

The following have reviewed the text and have provided valuable input:

Mike Awwadi, DeVry University, New Brunswick, New Jersey

Mike Beaver, University of Rio Grande, School of Technology, Rio Grande, Ohio

Joe Gryniuk, Lake Washington Technical College

David Holding, DeVry University, Kansas City, Missouri

Bill Liu, DeVry University, Fremont, California

Predrag Pesikan, DeVry University, Scarborough, Ontario

Gilbert Seah, DeVry University, Mississauga, Ontario

I would also like to thank my father, Gordon F. Snyder Sr., to whom this book is dedicated and who supplied many of the antique telephones and other items photographed for this text. Especially, I would like to thank my wife Diane and children Eva and Gabby for their help, understanding, and support.

Gordon F. Snyder Jr.

About the Author

Gordon F. Snyder Jr. is Executive Director and Principal Investigator for the National Center for Telecommunications Technologies (NCTT: http://www.nctt.org) at Springfield Technical Community College (STCC), where he also serves as project director of the *Microsoft Working Connections* grant program and manages curriculum development for networking. He has taught in the telecommunications, electronics systems, computer systems, and laser electro-optics departments at STCC since 1984, and co-chaired those departments from 1990 to 1999. He helped develop the *Verizon Next Step* program and now serves as the New England telecommunications curriculum coordinator for the Verizon Next Step program. He was an adjunct instructor in the bioengineering department at Western New England College and is the author of two other engineering textbooks. He has extensive consulting experience in the field of communications and LAN/WAN design. He serves on several local and national boards, including the Microsoft Community & Technical College Advisory Council, the Massachusetts Telecommunications Council, and the National Skill Standards Board (NSSB) Information and Communications Technology (ICT) Voluntary Partnership representing the telecommunications, computer, and information industry sector. In 2001, he was selected as one of the top 15 technology faculty in the United States by the American Association of Community Colleges and Microsoft Corporation.

Snyder received dual bachelor of science degrees in microbiology and medical technology from the University of Massachusetts Amherst, and the master of science in electrical engineering from Western New England College. In his spare time he enjoys most outdoor sports including fishing for striped bass on Cape Cod with family and friends.

An Introduction and Brief History

Introduction

During most of the 20th century, the telecommunications industry consisted of private companies subject to strict **Federal Communications Commission** (**FCC:** http://www.fcc.gov) regulation. Over time, the network was constructed and fine-tuned to allow for the relatively narrow voice range of frequencies, commonly referred to as **bandwidth.** With the emergence of the **World Wide Web (WWW)** in the mid-1990s, the bandwidth of this voice-tuned legacy network came to be no longer sufficient; and, as a result, the infrastructure continues to change at a very rapid pace.

1.1 History of the Telephone System

Before we can talk about the latest and greatest technology used by today's emerging high-bandwidth networks, we need to understand the current telecommunications infrastructure. The FCC chairman William E. Kennard places the magnitude of change in perspective in a speech he made on August 26, 1999. He said "We are standing at the threshold of a new century, a century that promises to be as revolutionary in the technology that affects our daily lives and the future of our country as the inventions that so profoundly shaped the past 100 years." Following is a short history of some of the major breakthroughs in telecommunications.

1876

- Alexander Graham Bell and Elisha Gray, another inventor competing with Bell, are both scrambling to get their voice-transmission inventions patented.

February 14, 1876

- On this day, Alexander Graham Bell's father-in-law, Gardinar Hubbard, delivers a patent application from Bell to the U.S. Patent Office for a device that transmits voice frequencies across wires.
- Approximately 3 hours later on the same day, Elisha Gray files a **caveat** (a formal notice requesting postponement of Bell's application) with the U.S. Patent Office describing a device Gray had hoped to patent that also transmits voice frequencies across wires.

March 10, 1876

- Alexander Graham Bell and Thomas A. Watson demonstrate a working telephone system but not without controversy. When Bell's original patent and Gray's caveat, both filed on February 14, were reviewed, it was determined that the device Bell described would not have worked while Gray's would have. It was speculated that Bell had copied parts of Gray's design. In Gray's caveat, he had detailed the use of a variable-resistance transmitter that was used to produce a transmitter signal ro-

FIGURE 1-1 Variable-resistance transformer.

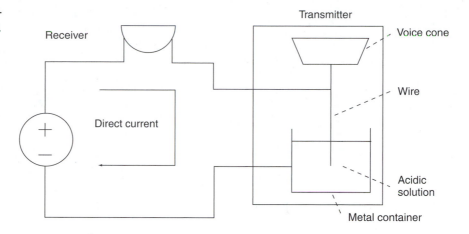

bust enough for the receiver to hear. Bell had been struggling to solve this same problem. In Bell's patent application, he made what appeared to be a last-minute handwritten notation about the use of a variable-resistance transmitter. People speculated that Bell had found out about Gray's caveat and had learned of Gray's use of a variable-resistance transmitter and, at the last minute before filing, had made a note on the patent application about using the new transmitter.

- The variable-resistance transmitter demonstrated by Bell on March 10, 1876, used a voice cone attached to a diaphragm. Also attached to the diaphragm was a wire that was immersed in a metal container of acidic solution, as illustrated in figure 1-1. The user talked into the voice cone, voice sound waves caused the diaphragm to vibrate, and the wire moved up and down in the acidic solution. As the wire moved up and down in the solution, the resistance between the wire and the metal container changed causing the direct current (DC) to vary in proportion to the variation in sound waves.

- The controversy between Bell and Gray led to years of litigation to the level of the U.S. Supreme Court where a split decision gave Bell the patent for the telephone entitled "Improvements in Telegraphy."

1877

- It takes a little over a year for Bell to acquire the Bell Telephone Company and to convince his father-in-law, Gardinar Hubbard, to finance the company and fund the building of the voice network infrastructure.

- Bell trys to sell the company to Western Telegraph.

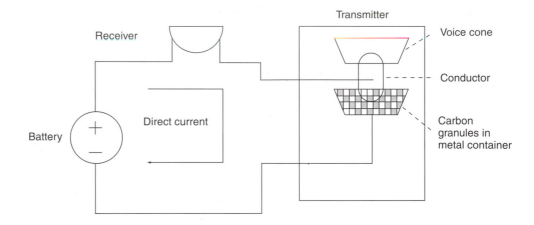

FIGURE 1-2
Hummings' carbon transformer.

1878

- Bell sets up the first operator switching exchange.

- Western Union Telegraph Company decides to use its existing national telegraph wire network to set up its own telephone company. Bell quickly sues Western Union; and Western Union settles out of court, selling its network to Bell.

- In England, Henry Hummings gets a British patent for a variable-resistance telephone transmitter that uses finely ground carbon. The carbon transmitter solved many of the early problems that Bell had trying to use liquid and electromagnetic transmitters. The carbon transmitter also used a voice cone attached to a diaphragm, as illustrated in figure 1-2 and in figure 1-3. The diaphragm, which was attached to a conductor, vibrated with sound waves and caused the closed container of ground carbon to compress and uncompress, changing resistance in the same way the liquid transmitters did.

1885

- American Telephone and Telegraph Company (AT&T: http://www.att.com) is formed by the Bell company to provide long-distance telephone service, connecting small Bell regional telephone franchises.

- AT&T buys Henry Hummings' ground carbon variable-resistance telephone transmitter patent rights.

1886

- Thomas Edison modifies Henry Hummings' finely ground carbon transmitter by using larger carbon granules. The larger granules created more current paths with sound wave compression and, therefore, allowed

FIGURE 1-3 Early Western Electric carbon transmitter.

FIGURE 1-4 Very early Western Electric wall phone reproduction with some authentic parts.

FIGURE 1-5 Early Western Electric wall phone reproduction with some authentic parts.

more current to flow in conjunction with the compression. The larger granules also did not pack as tightly over time as the finely ground carbon in Hummings' transmitter. When they did pack, lightly hitting the transmitter on a hard surface usually would loosen them up.

1899

- AT&T reorganizes, assuming the business and property of American Bell, and becomes the parent company of the Bell System.

1908

- Siemens (http://www.siemens.com) first tests **dialtone,** the telephone signal formed by the simultaneous transmission of a 330-hertz tone and a 440-hertz tone, on the public switched telephone network in a German city.

1913

- The Kingsbury Agreement, named after the AT&T vice president who wrote it, is delivered to the U.S. Justice Department. The agreement indicated that AT&T had ended any plans to build a complete telecommunications monopoly. AT&T agreed to sell its Western Union stock, to not buy any more independent telephone companies without first

FIGURE 1-6 Early Western Electric receiver.

getting government approval, and also to allow the independent telephone companies to connect to AT&T's long-distance lines.

1918

- AT&T patents an anti-sidetone solution for telephone receivers and transmitters. **Sidetone** is the sound of the telephone user's own voice heard in the telephone receiver. This technology allowed talkers to more easily adjust their voice volume when speaking into the telephone transmitter.

1930

- The bugs are worked out of the original AT&T patent, and anti-sidetone technology is added to telephones for consumer use.

1934

- Congress passes the Communications Act of 1934, creating the Federal Communications Commission (FCC: http://www.fcc.gov), to regulate the telecommunications industry.

FIGURE 1-7
Western Electric
Model 500
telephone.

1941

- Computers and the phone system are first combined to transmit data.

1947

- The transistor is invented.

1951

- The first transcontinental microwave system developed by AT&T begins operation. Between New York and San Francisco, the Bell System has 107 relay stations spaced approximately 30 miles apart.

1960

- The laser is invented.

1963

- AT&T introduces the digital or T-carrier technique and system. Prior to this, analog systems had been used.

1964

- Micro Communications Industries (MCI: http://www.mci.com) needs a medium for connecting their transmission lines to the radio towers used for wireless trucker communications between St. Louis and Chicago and asks AT&T for equal access in order to run wires to the

FIGURE 1-8 New England telephone technician, 1973.

towers. AT&T says no, resulting in a lawsuit against AT&T seeking equal access to long-distance networks.

1965

- AT&T launches the first commercial communications satellite, providing 240-two way telephone circuits.
- Western Electric builds and AT&T installs the first central office computerized switch, the No. 1 **electronic switching system (ESS),** in Succasunna, New Jersey.

1968

- The FCC allows non-Bell equipment to be legally attached to Bell System lines.

1969

- The Advanced Research Projects Agency (ARPA), part of the U.S. Department of Defense (http://www.defenselink.mil) develops **ARPANet,** a wide area network designed with redundant paths so that, in the event of war, communications can continue even though major pieces of the network may be destroyed.

1970

- First low-loss glass fiber is developed, and fiber-optic communication becomes practical.

1974

- The U.S. Justice Department investigates AT&T for violating the Sherman Antitrust Act.

FIGURE 1-9 AT&T technician, 1984.

1975

- Bill Gates and Paul Allen form Microsoft Corporation (http://www. microsoft.com).

1980

- ARPANet replacement begins with a separate new military network, the Defense Data Network, and **NSFNet,** a network of scientific and academic computers funded by the National Science Foundation (http://www.nsf.gov).

1982

- Southern Pacific Railway International (SPRINT: http://www.sprint.com) runs fiber-optic cable along its railroad beds and begins providing long-distance service.

1984

- Legal judgment is found against AT&T, causing the company to be divested, or broken up, into about 1000 small telephone companies. Prior to this date, AT&T provided the bulk of telecommunications in the United States. Suddenly, the industry was deregulated and, as a result, much more competitive. The age of divestiture had arrived.

1990

- Tim Berners-Lee of the European Council for Nuclear Research (http://www.w3.org/People/Berners-Lee) and a graduate of Oxford University, England, writes the software code for the first Internet browser-editor referred to as a **Web client.** People now want to "see" all the available resources on the Internet, commonly referred to as the World Wide Web, from their businesses and homes.
- A small network device company called Cisco Systems (http://www. cisco.com) goes public at $18.00 per share.

1992

- As a hobby, Linus Torvalds from Finland starts development of a free operating system called Linux for 386 and 486 (PCs).

1995

- Privatization begins as NSFNet is transferred to a consortium of commercial backbone providers (PSINet, UUNET, ANS/AOL, Sprint, MCI, and AGIS-Net99).
- In January, AT&T announces it will be splitting into three companies over the next fifteen months. These companies are: AT&T (http://www.att.com), to provide communication services; Lucent Technologies (http://www.lucent.com), to provide communications products; and NCR Corp. (http://www.ncr.com), the computer business.

1996

- The Telecommunications Act of 1996 (http://www.fcc.gov/telecom.html) is passed allowing Regional Bell Operating Companies more open competition for long-distance telephone traffic under certain circumstances.
- Lucent and NCR both become independent companies.
- NYNEX merges with Bell Atlantic to form the second largest telephone company in the United States.

1998

- Nortel Networks, Paradyne, and Rockwell announce **consumer asymmetric digital subscriber line (CADSL),** a technology that allows existing telephone lines to be used for both voice and high-speed data at the same time.
- V.90 56K modem standard is approved. **Modems** are devices used to transmit data over the existing voice telephone network.

2000

- Bell Atlantic and Vodafone AirTouch Plc announce Verizon Wireless (http://www.verizon.com).
- Bell Atlantic and GTE Corp. merge to form Verizon (http://www.verizon.com), an international wireline and wireless company with plans to offer services to over 40 countries.

- MediaOne and AT&T form AT&T Broadband (http://www.attbroadband. com).
- AT&T makes decision to restructure into four separate companies.
- V.92/V.44 analog modem standard approved by the **International Telecommunications Union** (**ITU:** http://www.itu.int).

2001

- America Online and Time Warner complete merger to create AOL Time Warner (AOL: http://www.aoltimewarner.com).

Summary

1. Alexander Graham Bell was in *direct competition* with Elisha Gray in the development of a device that transmits voice frequencies across wires.

2. On *March 10, 1876,* Alexander Graham Bell and Thomas A. Watson demonstrated a working telephone system.

3. The first transmitters, developed in the lab, used a *variable-resistance transformer* that required a cup of acidic solution be used.

4. *Western Union Electric Company,* which owned the telegraph network in the United States, sold its network to Alexander Graham Bell in 1878.

5. The first *operator exchange* was set up in 1878.

6. *AT&T* was first formed in 1885.

7. *Dialtone* was first tested on the voice network by Siemens in Germany in 1908.

8. The *FCC* was created in 1934 to regulate the telecommunications industry in the United States.

9. The first transcontinental *microwave system* was developed by AT&T in 1951.

10. The first *commercial communications satellite* was launched by AT&T in 1965.

11. The *Advanced Research Projects Agency (ARPA)* developed ARPANet in 1969 for defense communications.

12. AT&T was legally *divested,* or broken up, into about 1000 small telephone companies in 1984.

13. In 1990, *Tim Berners-Lee* wrote the code for the first Web client.

14. In 1995, NSFNet is privatized and becomes the modern *Internet.*

Discussion Questions

1. Using the Internet, research the role of the FCC at http://www.fcc.gov.

2. Using the Internet, research the role of the ITU at http://www.itu.int.

3. Contrast the FCC and the ITU using your research in questions 1 and 2.

4. Research the Western Union company on the Internet. What does the company do now?

5. Follow links listed in the chapter and locate some pictures of Tim Berners-Lee's first Web browser. Does it look much different than modern browsers? What type of computer did Berners-Lee use to develop the first browser?

6. Research the Telecommunications Act of 1996. Who was the judge responsible for the ruling?

7. Is cable modem and/or ADSL available in your neighborhood? Do some research and determine how much it will cost per month for each available service.

8. Talk with some older people you know (parents, grandparents, friends) and ask them if they remember how much long-distance service cost per minute prior to 1984. How does it compare to long-distance rates today?

The Basic Telephone Set

2

Introduction

The basic telephone set, connected to the telephone network we all very comfortably use, has four basic functions:

1. To provide a *signal* to the telephone company that a call is to be made or that a call is complete

2. To provide the telephone company with the *number* the caller wishes to call.

3. To provide a way for the telephone company to *indicate* that a call is coming in or ringing

4. To *convert* voice frequencies to electrical signals that can be transmitted at the transmitter and convert those electrical signals back to voice frequencies at the receiver.

The Federal Communications Commission (FCC: http://www.fcc.gov), the U.S. organization that sets telecommunications standards, has set standards for the aforementioned features; all manufacturers selling telephones in this country must match these standards or the phone will not work properly. In addition, many modern telephones also come with features like speed dial, redial, memory, caller ID, voice mail, and so on. These are all additional features that are not necessary to make or receive calls.

2.1 Network Connection Basics

Before we get inside the telephone, let us look at how the telephone is connected to the telephone network. The **local-exchange carrier (LEC)** is the telephone company responsible for the final telephone network connection coming into a home or business. The LEC can be a regional telephone company referred to as a **Regional Bell Operating Company (RBOC)** or a competitor. This connection is made through a **central office (CO)** or **central exchange (CE).** The central office (CO) or central exchange (CE) is a building that is maintained by the LEC and that houses the telephone switching equipment connected to the lines running in from the local homes and businesses. The wire connection between the telephone being used in the residence or business and the local-exchange carrier (LEC) central office (CO) is called the **subscriber loop** or the **local loop.** The local loop is commonly a pair of twisted copper wires that come into to the home or business and is shown in figure 2-1.

Tip and Ring

The terms **tip** and **ring** are used to designate each of the local loop pair wires. This designation comes from the days when human operators used a manual

FIGURE 2-1 The local loop.

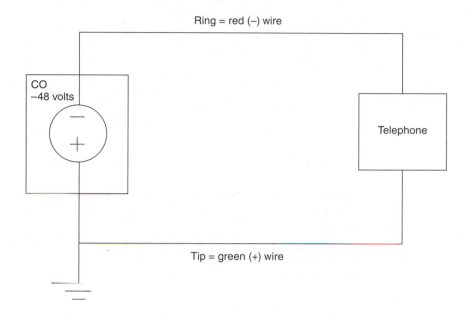

Ring = red (–) wire

CO
–48 volts

Telephone

Tip = green (+) wire

FIGURE 2-2 Typical tip-and-ring switchboard jacks.

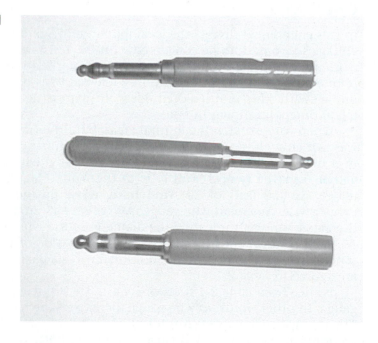

switchboard and tip-and-ring connectors called *jacks* to connect one party's local loop to another party's local loop. Some typical tip-and-ring switchboard jacks are shown in figure 2-2, and jack details are shown in figure 2-3.

Each loop was connected to a jack similar to the ones shown in figure 2-2. Callers would pick up a receiver and turn a magneto generator handle that

FIGURE 2-3 Tip-and-ring switchboard jack details.

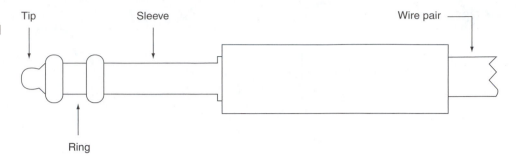

would generate a current on the loop and cause a bell to ring on the switchboard. Figure 2-4 and figure 2-5 show two different hand crank magneto telephones.

The operator would attach to the caller's line using a tip-and-ring plug and ask callers who they wished to call. The operator would then build a circuit using tip-and-ring switchboard jacks and wires connecting the two callers.

Open Wire

In this chapter, we discuss the local loop in terms of direct current (DC) loop voltage so we need to be concerned with DC resistance. The early voice network used uninsulated thick copper wire strung from telephone pole to telephone pole. This wire was commonly referred to as **open wire.** Most are familiar with the multicolored glass insulators used on the open-wire telephone poles shown in figure 2-6.

Open wire worked fine in rural areas; but, in more congested areas, there was not enough room on the poles for the number of wires needed to connect everyone. The solution was to use insulated wire and bundle wires together. Today, a typical local loop in the United States is 3 miles (approximately 18,000 feet) of 22 **American wire gauge (AWG)** twisted pair copper wire. American wire gauge (AWG) is a U.S. numbering standard with the gauge number indicating the diameter of the wire.

A 22 AWG wire carries a resistance specification of 19 ohms (Ω) per thousand feet at 77° F (25°C). Some additional wire gauge resistance specifications are shown in table 2-1.

Notice from table 2-1 that wire resistance is directly proportional to wire gauge. In other words, as wire gauge increases, wire resistance also increases. The wire gauge scale is logarithmically based and is an indication of the cross-sectional area of a wire. A 2-gauge step-down in gauge corresponds to a doubling of the cross-sectional area, while a 2-gauge step-up in gauge corresponds to a halving of the cross-sectional area. This means a 19-gauge wire has twice the cross-sectional area of a 22-gauge wire and will have half the resistance.

FIGURE 2-4
Western Electric Model 202C telephone with Model 315H magneto (*top*). Close-ups are shown in the middle and bottom photos.

FIGURE 2-5 Federal set with built-in magneto.

FIGURE 2-6 Typical early open-wire glass insulators.

TABLE 2-1 Sample wire gauge resistance specifications.

Wire gauge	Ω per 1000 feet
19	9.5
22	19.0
24	30.2
26	48.0

FIGURE 2-7 Local loop transmission line showing lumped resistance.

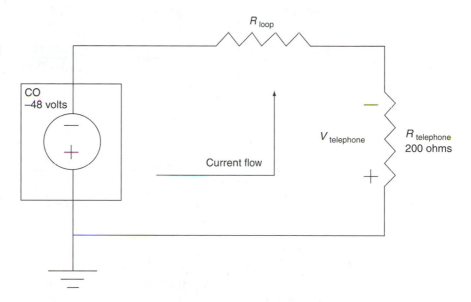

On the local loop, we are very concerned with DC wire resistance. The object of any transmission system is to deliver as much input (CO) power as possible to the output device (telephone at the end of the loop). Figure 2-7 shows a local loop transmission line with all loop resistance lumped into one single resistance value, R_{loop}.

To calculate voltage drop caused by the wire resistance in the local loop, we can use a simple voltage divider.

$$V_{telephone} = \left(\frac{R_{telephone}}{R_{loop} + R_{telephone}} \right) -48 \text{ V}$$

If loop resistance is too high, there will not be enough voltage—and corresponding current—for the phone to work properly. All voltage will have been dropped across the local loop.

●─**EXAMPLE 2.1**

●─**Problem**

A local loop is made of 22 AWG wire and is 18,000 feet long. The connected telephone has an internal DC resistance of 200 ohms. Calculate the amount of DC voltage the telephone will actually see.

●─**Solution**

$$R_{loop} = \text{loop DC resistance}$$

$$= (18,000 \text{ ft})(19 \ \Omega \ / \ 1000 \text{ ft for 22 AWG copper wire})$$

$$= 342 \ \Omega$$

$$V_{telephone} = \left(\frac{R_{telephone}}{R_{loop} + R_{telephone}} \right) - 48 \text{ V} = \left(\frac{200 \ \Omega}{342 \ \Omega + 200 \ \Omega} \right) - 48 \text{ V} = -17.7 \text{ V}$$

Later in this chapter, we show that this is enough voltage for the telephone to function properly.

Notice the CO provides a negative (–) 48-volt DC to each connected telephone. Since 1894, COs in the United States have had large battery banks that provide this DC voltage. This is why corded telephones (not cordless) do not require AC power and why you can still make calls on them during power outages; the telephones are powered by batteries in the CO. The U.S. Rural Electrification Administration (REA) recommends that the battery provided should have the capacity to maintain the CO load for a period of 8 hours. Systems that are equipped with emergency generators are allowed to reduce the 8 hours to a 2-hour reserve time. Central office equipment also depends on this –48-volt DC supply. Central office digital equipment is especially susceptible, and performance of digital equipment at voltages less than 44 volts becomes unpredictable.

Some specialized equipment (examples: radio carrier equipment and pay phone coin-collect circuits) require voltage levels different from the –48-volt DC supplied by the CO. Common values are ±24-volt DC, ±48-volt DC, and ±130-volt DC. In these cases, DC to DC converters are used to increase or decrease the –48 volts and produce these different voltages. Knowledge of the power required by these DC-DC converters is critical at the CO level. The power requirements of all devices running off CO batteries must be included in the total load CO DC power system calculations to determine battery size and number along with emergency power generator specifications. Using a DC voltage also eliminates any AC hum on the lines caused by ground differentials.

FIGURE 2-8 Positive CO voltage with different ground potentials.

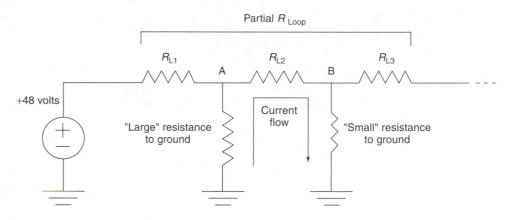

Why the Negative Voltage?

The CO voltage is negative for *electrolytic corrosion* reasons. We understand that the voltage supplied to the telephone is delivered by a pair of wires and provides a complete circuit to the local loop. We also understand that the local loop wire has an amount of resistance determined by wire gauge, wire distance, and temperature. In a long loop, this wire resistance can result in significant voltage differences in potential between the CO and the connected telephone.

What If the Voltage Was Not Negative?

Let us assume that the positive terminal of the CO battery is attached to the nonground side, as shown in figure 2-8. If one end of the loop has a better connection to earth than the other, the other end will be at a small amount of voltage above earth. Additional points along the length of wire also may be making ground contact. If we take a piece of the local loop and spread the loop resistance out instead of lumping it all together, we can see how these ground contact points can cause problems. Let us look a little closer in figure 2-9 and add a resistance indicating ground.

In figure 2-9, point B is making better contact to ground than point A. Maybe insulation around a wire is worn or a splice has gotten wet. As a result, point A will be at a voltage potential higher than point B. In this case, some of the current will tend to leave the loop at the better-connected point A and find its way back to complete the circuit through the earth. These resulting currents are called *leakage currents,* and these currents cause corrosion. As current passes through the metal loop wires and connectors, it slowly removes the metal in a process called *electrolytic corrosion.* If this happens over a long time period, serious corrosion problems will occur. Wire systems are

FIGURE 2-9 Positive CO voltage with different ground potentials.

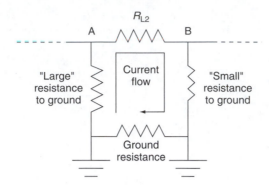

FIGURE 2-10 CO diagram with –48-volt DC battery supply.

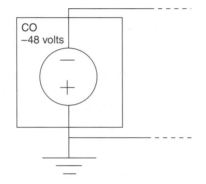

not the only things that can corrode this way; bridges, pipe systems, metal foundations, electric train rail systems—just about anything that can conduct electricity and is either attached or close to ground has the potential to electrolytically corrode.

The Solution: Reverse the Battery Polarity

To solve the problem, telephone companies reverse the battery polarity in the CO, as illustrated in figure 2-10. By reversing the polarity in the CO, telephone companies have provided a less-resistive path through the negative or ring wire than through ground for current to flow back to the CO and complete the local loop circuit.

Lines longer than 3 miles may have amplifiers called *loop extenders* and *voice-frequency repeaters* attached. These devices extend the length of the local loop. They are powered by the –48-volt DC from the CO and incorporate a voltage booster and a voice amplifier that ensure a consistent loop current.

FIGURE 2-11
Western Electric Model 211C telephone attached to early switchboard showing switchhook detail.

2.2 The Telephone Set

Now that we know a little bit about the local loop, let us look at each telephone function in a little more detail.

Telephone Set Function 1: To Provide a Signal That a Call Has Started or a Call Is Finished

The **switchhook** is the switch on a telephone handset that closes when a receiver is picked up causing current to flow from the CO switch on the local loop. It gets its name from the old telephones that had a hook on the side, as shown in figure 2-11. Today's telephones commonly use buttons to open and close two switchhooks, as shown in figure 2-12.

FIGURE 2-12 Local loop and switch-hooks (S1 and S2).

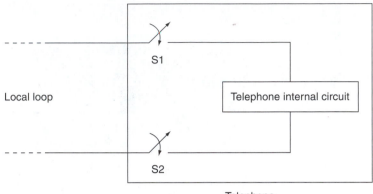

Telephone

According to telephone company specifications, individual telephone set DC resistance should be 200 ohms; but, in reality, most telephones range between 150 and 1000 ohms of DC resistance. When a user picks up a connected telephone handset to make a call, the switchhooks (S1 and S2 in figure 2-12) close (off-hook condition) and the local loop circuit is complete.

Let us look at an example and calculate the current in a typical local loop.

●—EXAMPLE 2.2

●—Problem
We can look at the same local loop in example 2.1 of this chapter. The loop is approximately 3 miles (18,000 feet) long and has a telephone with an internal DC resistance of 200 ohms attached to it. We have calculated the telephone voltage to be –17.7 volts. Calculate the amount of loop current that will flow when the handset is picked up on the phone.

●—Solution

Telephone voltage = –17.7 V

Telephone resistance = 200 Ω

We can use these values and Ohm's law $\left(Current = \dfrac{voltage}{resistance} \right)$ to calculate the current on the entire loop:

$$Loop\ DC = \frac{telephone\ voltage}{telephone\ resistance} = \frac{-17.7\ V}{200\ \Omega} = -88.5\ mA$$

FIGURE 2-13
Dialtone generation.

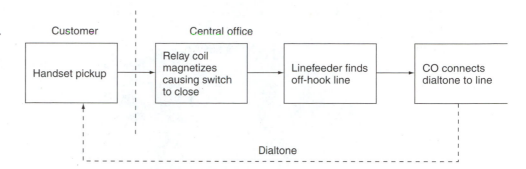

Example 2.2 shows a good approximation of local loop current flow when a handset is picked up. Direct current ranging between 20 and 120 milliamperes flows on the pair of wires connecting the telephone to the CO. This current flow causes a **relay coil,** a device used in a telephone company switch to determine that a customer wishes to make a call, to magnetize and its contacts close. In the CO, current flows through a relay coil attached to the local loop wire pair. The coil energizes, its contacts close, and the CO switch knows a phone is off-hook somewhere. A line feeder in the CO switch looks for the off-hook signal, finds it, and sets up a connection. In the CO switch, a dialtone generator is connected to the line so the caller knows a number can be dialed. This process is illustrated in figure 2-13.

Dialtone is a signal formed by the simultaneous transmission of a 330-hertz tone and a 440-hertz tone. After the first number is dialed by the caller, the dialtone generator is shut off.

The switchhook also is used on some expanded-feature phones for additional signaling such as call transfer and call hold.

Telephone Set Function 2: To Provide the Number Being Called to the Connected Telephone Switch

Two methods are used currently to provide numbers to the telephone company—pulse or **rotary dial service** and **dual-tone-multifrequency (DTMF) signaling.**

Pulse or Rotary Dial Service

In the past, when a handset was lifted, the caller did not hear dialtone; the caller heard an operator asking for the number the caller wanted to dial. As the number of telephones grew, telephone companies projected that hundreds of thousands of new operators would be needed so rotary dials (see figure 2-14) and automated switching were added to the **public switched telephone network (PSTN)** or **plain old telephone system (POTS).**

FIGURE 2-14
Western Electric
Model 302 with
rotary dialer.

Rotary dial service was invented to eliminate operators. It uses dial pulsing to automate the switching required to get from a caller to a receiver. The rotary dial generates pulses on the local loop by opening and closing an electric switch when the dial is rotated and released. Each pulse opens the loop and interrupts the local loop current flow of 20 to 120 milliamperes, resulting in a series of current pulses on the local loop. This process is referred to as **out-pulsing;** with pulse output at approximately 10 pulses per second. Each pulse is actually an interruption in current flow on the loop and is .05 second with a .05-second pause between pulses. Each number on the dial corresponds to the number of pulses produced for that number. For example, dialing the number *4* produces four pulses and takes a total of .4 second (.05 second × 8 = .4 second). This is illustrated in figure 2-15; and, as you can see, rotary dialers are slow when compared to modern telephones.

●—EXAMPLE 2.3

●—Problem
How long does it take to dial the single number *9* on a mechanical rotary phone?

●—Solution
Dialing the number *9* produces:

.05-second pulse, .05-second pause, .05-second pulse, .05-second pause, .05-second pulse, .05-second pause, .05-second pulse, .05-second pause, .05-second pulse, .05-second pause, .05-second pulse, .05-second pause, .05-second pulse, .05-second pause, .05-second pulse, .05-second pause, .05-second pulse, .05-second pause.

.05 second × 18 = .9 second

FIGURE 2-15
Telephone rotary dial timing diagram of the number *4*.

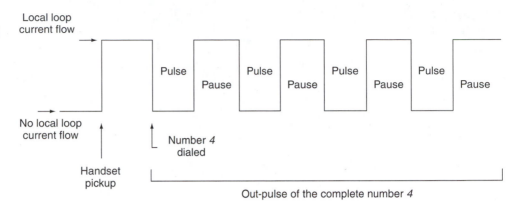

As telephone-manufacturing technology developed, the rotary dials on many phones were replaced with push-button keypads. These keypads use an electronic circuit to generate the pulses, not a mechanical rotary dial. This type of dial is equipped with a buffer that stores numbers as they are keyed. The buffer then out-pulses the numbers with the proper timing intervals.

●—EXAMPLE 2.4

●—Problem
How long does it take to dial the single number *9* on a pulse dial push-button keypad?

●—Solution   Dialing the number *9* still produces:

.05-second pulse, .05-second pause, .05-second pulse, .05-second pause, .05-second pulse, .05-second pause, .05-second pulse, .05-second pause, .05-second pulse, .05-second pause, .05-second pulse, .05-second pause, .05-second pulse, .05-second pause, .05-second pulse, .05-second pause, .05-second pulse, .05-second pause.

.05 second × 18 = .9 second

Dual-Tone-Multifrequency Signaling Service

The most commonly used method for inputting a number in the United States and Europe is now the *dual-tone-multifrequency (DTMF)* signaling method. DTMF telephones are also commonly known as Touch-Tone™

FIGURE 2-16
Telephone DTMF
keypad.

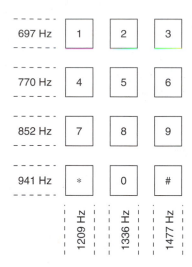

telephones. These phones also use numerical keypads but offer an even faster way to signal the number to call by sending tones on the telephone line. The DTMF signaling method uses a 12-button telephone keypad; and, when a button is pressed on the keypad, an electric contact is closed and two tones are generated at the indicated frequencies. Figure 2-16 illustrates a DTMF telephone keypad.

The frequencies used are illustrated in the keypad diagram. For example, looking at figure 2-16, notice that when the number *8* is pressed, the 852-hertz and 1336-hertz frequencies are combined to form the number *8* tone.

The tones must be 50 milliseconds long with a 50-millisecond pause between each. DTMF phones offer much more rapid dialing of numbers than rotary pulse methods.

●—EXAMPLE 2.5

●—Problem
How long does it take to dial the single number *9* on a DTMF keypad?

●—Solution   Dialing the number *9* produces:

.05-second tone, .05-second pause.

.05 second × 2 = .1 second

FIGURE 2-17 Ring or alerting signal.

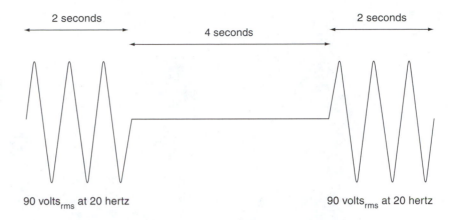

Telephone Set Function 3: To Indicate There Is an Incoming Call

When the user dials the phone, each sequenced number is stored in the CO computerized switch at the LEC CO and analyzed. The first three digits determine if the call is *local* or *long distance*. If the call is local, the switch determines if it can complete the call itself or if the call needs to be forwarded to another local LEC CO that handles that telephone number. If the call is long distance, the call needs to be forwarded to the customer's long-distance carrier. Once the call destination is determined, a switch on the receiving end sends a repeating 90-volt$_{rms}$ 20-hertz ringing signal (on for 2 seconds with a 4-second pause) called a *ring* or *alerting signal* to the receiving phone causing it to ring. Figure 2-17 indicates this ringing signal. Notice that the ringing system is an alternating current (ac) voltage in root mean square (rms) units. Also note that the ringing signal has an inaudible frequency of 20 hertz; this is why different phones can have different ring styles.

At the same time, a ring-back signal that is a mix of 440 hertz and 480 hertz is sent back to the caller. This signal is on for 2 seconds and off for 4 seconds and indicates that the phone being dialed is ringing. When the receiver picks up the handset, the telephone goes *off-hook*. The switchhook on the receiving phone closes, current flows, and the CO switches turn off the ringing signals.

Telephone Set Function 4: To Convert Voice to Electrical Frequencies and Electrical Frequencies to Voice

A telephone converts voice frequencies to electric signals and electric signals back to voice frequencies using basic microphone transmitter and speaker theory and application. Figure 2-18 shows some early telephone transmitters and receivers.

FIGURE 2-18 Early Western Electric transmitters and receivers.

FIGURE 2-19 Telephone carbon granule transmitter.

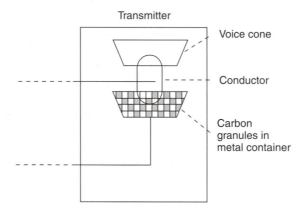

Transmitters

A telephone **transmitter** in the telephone handset converts voice into electrical frequencies. The most common transmitter still in use today is the carbon granule transmitter, as illustrated in figure 2-19.

Sound travels in waves, and when you talk into a handset some of these waves enter the mouthpiece and cause a diaphragm in the transmitter microphone to vibrate against a container filled with carbon granules. The vibrations cause the granules to pack and unpack. Electrical contact is made with the granule container, and voltage is applied across the contacts. Varying

FIGURE 2-20
Simple speaker
diagram.

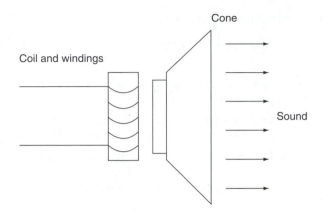

amounts of resistance caused by the carbon granules in the microphone cause varying amounts of current to flow is an electrical representation of the voice coming into the transmitter.

In addition to carbon granule transmitters, many modern telephones use *dynamic transmitters* that function by moving a coil of wire inside a magnetic field to produce an electric current in response to sound waves; or *electret transmitters*, also known as *condenser microphones*, which use a capacitor for a transducer and generally contain an amplifier circuit.

Receivers

The telephone handset **receiver** is just a simple speaker built into the telephone handset. It performs the opposite function of the transmitter in that it converts an electrical signal to sound waves, as illustrated in figure 2-20. The signal flows through a coil in the back of the speaker, and the magnetic field formed by the coil changes as the changing current flows through the coil. This changing magnetic field causes a cone in the speaker to vibrate. These vibrations create air pressure waves forming sound.

2.3 Additional Telephone System Features

A few other common telephone system features that we are all used to having and relying on are additional handset signals, **caller ID,** and **preferred interexchange carrier (PIC).**

Some Common Handset Signals

We are all used to hearing these additional common signals coming from our telephone:

- *Line busy signal.* 480-hertz and 630-hertz tones on for .5 second and off for .5 second, then repeat.
- *Block signal.* 480-hertz and 620-hertz tones on for .2 second and off for .3 second, then repeat. This signal is often referred to as *fast busy.*
- *Off-hook.* 1400-hertz, 2060-hertz, 2450-hertz, and 2600-hertz tones on for .1 second and off for .1 second, then repeat with a duration of 40 seconds. This signal is designed to be heard from across a room and is very loud.

Caller ID

Caller ID, also referred to as *caller identification technology,* identifies who is calling to the receiver and was introduced in New Jersey by Bell Atlantic in 1987. Caller ID became widely available in the United States with the implementation of a switching technology called **Signaling System 7 (SS7).** The *SS7* network has been designed to only transmit signals—like caller ID—and not voice. The SS7 network works in conjunction with but is separate from the voice network and is discussed in detail in chapter 7. The SS7 network is necessary to provide **custom local area signaling services (CLASS).** These services are supplemental telephone services and include caller ID, call return, repeat dialing, priority ringing, select call forwarding, call trace, and call blocking.

To transmit caller ID information, the SS7 network sends the telephone numbers of the caller and the recipient in the form of a signal to a transfer point before the call arrives at the receiving end. The caller ID information, also referred to as the **calling party number (CPN)** field, is placed .5 second after the first ring between the first and second rings of a telephone call by the SS7 network in the transmitting CO and contains caller identification information. This is illustrated in figure 2-21.

As the call travels over the caller-to-receiver route, it may pass through network elements that have not yet been upgraded to SS7. If this happens, the CPN information will disappear and the receiver will not be able to identify who is calling.

The CPN information is delivered in *frequency-modulated digital data format.* Simply put, this format is a series of *bits* (binary digits), or *1s* and *0s,* that are transmitted sequentially. The bits are organized into groups of eight, with each eight-bit group referred to as a *byte.* Each eight-bit combination, or byte, represents a letter of the alphabet, number, or special character. In the CPN field, a *1* is represented by a 1200-hertz tone and a *0* is represented by a 2200-hertz tone. The layout of the CPN field is shown in more detail in figure 2-22.

FIGURE 2-21 Caller ID or calling party number (CPN) field.

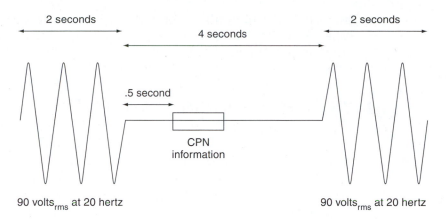

FIGURE 2-22 Calling party number (CPN) field detail.

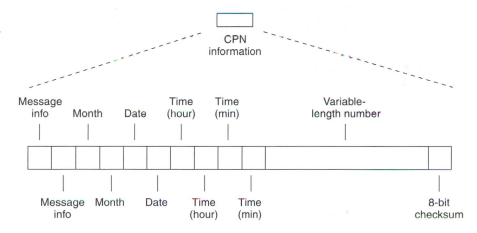

The message information fields include the type of information that is included in the CPN field and how long (how many bytes) the CPN field is. The variable-length number field can include name and/or address information. Caller ID requires the use of a telephone capable of displaying caller ID information or a display box attached to a non-caller-ID-display phone. The receiver must wait for the time between the first and second rings to see caller ID information. If the receiver picks up the phone before the CPN information is received, no caller ID information is displayed. This easy accessibility to caller numbers and other information concerns many people and has created a privacy debate. The constitutionality of caller ID has been repeatedly challenged in court. People have three major concerns: the right to be left alone, the right to be free from unreasonable searches or seizures, and the right to not be subjected to unreasonable government intrusions. *Call blocking* has originated as a solution to this privacy problem. Call blocking allows callers to prevent their numbers from being displayed on the receiver's caller ID unit. Caller ID can be blocked in two ways. *Per-line blocking* blocks the caller's ID from all connected telephones. *Per-call blocking* requires the caller to dial *67 prior to dialing any call for which the caller wishes to be anonymous. When a CPN is blocked, the telephone company replaces the CPN information with *Private Name Private Number* or something similar.

Preferred Interexchange Carrier (PIC)

Since the 1976 MCI ruling, AT&T has been required to open the long-distance market to other long-distance providers. Prior to this, all long-distance traffic in the United States was handled by AT&T and users would just dial a *1* to connect to an AT&T long-distance trunk. As other long-distance carriers entered the market, AT&T had a big advantage. Customers were already used to dialing a *1* for long distance, and placing a long-distance call to anywhere in the United States involved dialing a minimum number of numbers—only 11. This included *1,* the area code, and the seven-digit number. Customers who wanted to use other long-distance carriers had to dial 25 numbers to make a long-distance call. These calls required that an 800 number be called initially (11 numbers), a 4-number personal identification number (PIN), the area code, and the 7-digit number. In 1987, a method called **Feature Group D** was implemented to automatically pass calls to the customers' preferred long-distance carrier using a *preferred tnterexchange carrier (PIC)* number. The PIC is the long-distance carrier selected by the customer. Customers are required to select a preferred carrier, and the preferred carrier information is added to the local switch database to which the customers are connected. Feature Group D also allows a customer to bypass the preferred PIC by dialing a *101XXXX* number and to use another long-distance carrier. These *101XXXX* are commonly referred to as *dial-around service numbers.*

Summary

1. The basic telephone set provides four basic functions:
   a. Provides a *signal* to the telephone company that a call is to be made
   b. Provides the telephone company with the *number* caller wishes to call
   c. Provides a way for the telephone company to *indicate* a call is coming
   d. *Converts* voice frequencies to electric signals that can be transmitted and then converts them back to voice frequencies at the receiver

2. The *FCC* has set standards for the basic telephone set to ensure compatibility with different telephone manufacturers.

3. The telephone company *central office (CO)* is a building maintained by the telephone company that houses telephone switching and trunking equipment.

4. The connection between a telephone and the telephone company CO is called the *subscriber* or *local loop.*

5. Early telephone local loops were made with uninsulated thick copper wire referred to as *open wire.*

6. Local loop *electrical resistance* needs to be kept low to ensure that there is enough voltage and current for connected telephones to work properly.

7. The telephone company provides *–48 volts DC* to each connected telephone using battery banks in the COs.

8. The telephone-company-provided –48 volts DC is made negative to suppress *electrolytic corrosion.*

9. The telephone set *switchhook* is used to open and close the local loop connection.

10. Rotary telephones use *dial* or *out-pulsing* to automate switching required to complete a telephone call.

11. Dual-tone-multifrequency (DTMF) telephones use *frequency tones* to automate switching required to complete a telephone call.

12. Telephone voice *transmitters* typically use carbon granule or condenser microphones.

13. Telephone voice *receivers* are simple speakers.

14. Additional telephone handset audible signals, such as *line busy, block,* and *off-hook,* are made up of combinations of audible frequencies.

15. The *caller ID information* is included between the first and second signals of a telephone call.

16. *Preferred interexchange carrier (PIC)* numbers are included in the local switch customer database.

Review Questions

Section 2.1

1. The _____ sets standards for U.S. telephone set features to ensure compatibility between manufacturers.
   a. FCC
   b. FBI
   c. CIA
   d. NSA

2. Supplemental features like speed dial, caller ID, and telephone memory are now required to make or receive telephone calls. True/False

3. The connection between a telephone and the telephone company central office is commonly called the _____.
   a. remote loop
   b. local loop
   c. distant connection
   d. loopback connection

4. As copper wire thickness increases, the electrical resistance of the wire decreases. True/False

5. 19 AWG wire has a smaller cross-sectional area than 22 AWG wire. True/False

6. In the United States, telephone companies must provide a voltage of _____ to power telephones.
   a. +24 volts DC
   b. –24 volts DC
   c. +48 volts DC
   d. –48 volts DC

7. A typical local loop in the United States today is _____ feet.
   a. 60
   b. 50,000
   c. 18,000
   d. 240,000

8. If local loop resistance is too low, there will not be enough voltage and corresponding current for connected telephones to work properly. True/False

9. Corded telephones typically will work during an electrical power outage. True/False

10. _____ can be used to ensure consistent loop current on extremely long local loops.
    a. Loop inhibitors
    b. Loop references
    c. Loop filters
    d. Loop extenders

Section 2.2

11. Telephone set internal resistance typically ranges between _____ ohms of DC resistance.
    a. 1.5 and 10
    b. 15 and 100
    c. 150 and 1000
    d. 1.5K and 10K

12. The telephone set _____ is used to provide a signal to the telephone company that a call is to be made.
    a. cradle
    b. switchhook
    c. transmitter
    d. receiver

13. When current flows in the local loop, a _____ in the CO switch magnetizes causing current to flow on the loop.
    a. transistor
    b. diode
    c. relay coil
    d. capacitor

14. Dialtone is a signal formed by the simultaneous transmission of _____ hertz tones.
    a. 110- and 220-
    b. 220- and 330-

c. 330- and 440-
d. 440- and 550-

15. When a handset is picked up on a connected telephone, a DC value ranging between _____ milliamperes flows on the local loop wire pair.
    a. 2 and 12
    b. 20 and 120
    c. 40 and 240
    d. 50 and 350

16. Rotary handset dialers are faster than DTMF handset dialers. True/False

17. A rotary dialer _____ the local loop circuit with each pulse.
    a. opens
    b. closes
    c. conditions
    d. measures

18. DTMF is an acronym for _____.
    a. dual-tone multiple farad
    b. digital-time multifrequency
    c. digital-tone multifrequency
    d. dual-tone multifrequency

19. The first _____ digits of a telephone call that are dialed determine if the call is local or long distance.
    a. 2
    b. 3
    c. 7
    d. 10

20. The ringing signal generated by the telephone company and sent to the receiving telephone has a frequency of _____ hertz.
    a. 20
    b. 40
    c. 60
    d. 80

21. Microphones that function by moving a coil of wire inside a magnetic field are known as _____ transmitters.
    a. carbon granule
    b. dynamic
    c. electret
    d. string

22. Microphones that function using a capacitor for a transducer are known as _____ transmitters.
    a. carbon granule
    b. dynamic
    c. electret or condensor
    d. string

23. Microphones that function by moving a diaphragm against a container of carbon granules are known as _____ transmitters.
    a. carbon granule
    b. dynamic
    c. electret
    d. string

Section 2.3

24. A telephone line busy signal is a repeating combination of _____ signals that are on for .5 second and off for .5 second.
    a. 10-hertz and 20-hertz
    b. 200-hertz and 400-hertz
    c. 480-hertz and 630-hertz
    d. 480-hertz and 620-hertz

25. A telephone line block signal is a repeating combination of _____ signals that are on for .2 second and off for .3 second.
    a. 10-hertz and 20-hertz
    b. 200-hertz and 400-hertz
    c. 480-hertz and 630-hertz
    d. 480-hertz and 620-hertz

26. An off-hook signal will repeat for an infinite duration. True/False

27. _____ is necessary to provide caller ID services.
    a. SS7
    b. SST
    c. TCP/IP
    d. PIC

28. CLASS is an acronym for _____.
    a. custom local area signaling services
    b. customer local access signaling signature
    c. custom local area signaling signature
    d. customer local access signaling services

29. It is possible (depending on call path) for caller ID information to be lost between the caller and receiver telephones. True/False

30. Caller ID information can be blocked if the caller dials _____ prior to dialing any call.
    a. #67
    b. #69
    c. *67
    d. *69

31. A customer's _____ identifies the customer's preferred long-distance carrier.
    a. CLASS
    b. SS7
    c. LEC
    d. PIC

32. A customer's preferred long-distance carrier can be bypassed by dialing _____. This allows use of another long-distance carrier (where Xs indicate additional numbers).
    a. 1010XXX
    b. 101XXXX
    c. 10XXXXX
    d. 1XXXXXX

Discussion Questions

Section 2.1

1. Define the terms *tip* and *ring*. What is the voltage on the tip wire? What is the voltage on the ring wire? What are the typical wire colors used for each?

2. A local loop is made up of 19 AWG wire and is 7500 feet long. A connected telephone has an internal resistance of 550 ohms. Calculate the total loop resistance including both wire and telephone resistances.

3. For the local loop and telephone combination in question 2, calculate the amount of DC voltage the telephone will actually see assuming the CO supply is –48 volts DC.

4. For the local loop and telephone combination in question 2, calculate the amount of current that will flow through the local loop.

5. A local loop is made up of 22 AWG wire and is 7500 feet long. A connected telephone has an internal resistance of 550 ohms. Calculate the total loop resistance including both wire and telephone resistances.

6. For the local loop and telephone combination in question 5, calculate the amount of DC voltage the telephone will actually see assuming the CO supply is –48 volts DC.

7. For the local loop and telephone combination in question 5, calculate the amount of current that will flow through the local loop.

8. Which is a better local loop connection—the local loop using 19 AWG wire in problem 2 or the local loop using 22 AWG wire in problem 5? Explain why.

9. Explain why the COs provide –48 volts DC instead of +48 volts DC to the customer.

Section 2.2

10. A telephone set malfunctions and assumes an internal resistance of 10 kilohms. Will this telephone likely work when connected to the local loop? Why or why not?

11. How long does it take to dial the single number *5* on a mechanical rotary telephone?

12. How long does it take to dial the long-distance number *18004219556* on a mechanical rotary telephone?

13. How long does it take to dial the single number *5* on a DTMF keypad telephone?

14. How long does it take to dial the long-distance number *18004219556* on a DTMF keypad telephone?

15. The telephone companies all provide a 90-volt$_{rms}$ 20-hertz ringing signal to call-receiving telephones, but different telephone sets have different ring sounds and patterns. Explain how this happens.

16. Early carbon granule telephone transmitters were commonly *fixed* when they were not working properly by hitting them a few times against a hard surface. Why do you think this occasionally worked?

Section 2.3

17. Describe how Feature Group D allows customers to bypass their selected PIC.

The Local Loop and Beyond

Key Terms

American National Standards Institute (ANSI)

analog modem

asynchronous transfer mode (ATM)

ATM cell header

bipolar with 8-zero substitution (B8ZS)

bit robbing

bits per second (bps)

bridged tap

cell

CODEC

companding

cutoff frequency

dense wavelength division multiplexing (DWDM)

digital multiplexing

DS-0

DS-1

DS-2

DS-3

DS-4

DS-5

E-carrier

encoding

frame

frame relay

framing

frequency division multiplexing (FDM)

frequency multiplexing

group

high-level data link control (HDLC)

International Telecommunications Union (ITU)

jumbogroup

jumbogroup multiplex

loading coil

low-pass filter

masterframe

mastergroup

multiple subscriber line carrier system

multiplex

node

nyquist sampling theorem

optical carrier (OC)

overhead

packet assembler/disassembler (PAD)

packet-switched network

packet-switching exchange (PSE)

payload

permanent virtual circuit (PVC)

pulse amplitude modulation (PAM)

pulse code modulation (PCM)

quantization

quantizing noise

remote terminal (RT)

statistical time division multiplexing (STDM)

subscriber line carrier (SLC)-96

superframe

supergroup

synchronous optical network (SONET)

synchronous transport signal level (STS)

time division multiplexing (TDM)

wavelength

wavelength division multiplexing (WDM)

Introduction

Early in the development of the telephone system infrastructure, designers realized that everyday speech of humans lies in between 125 hertz and 8 kilo-hertz, with most voice centered between 400 and 600 hertz. With more studies, designers realized that humans can recognize and interpret voices if they stayed within this frequency range. As illustrated in figure 3-1, voice frequencies below 200 hertz and above 2 kilohertz play very little role in voice recognition; and, for this reason, the public switched telephone network (PSTN) has been tuned to a frequency range of approximately 300 hertz to 3300 hertz. Since the early 1900s, the infrastructure has been tuned to match these frequency requirements.

3.1 The Local Loop

The analog public switched telephone network (PSTN) or plain old telephone system (POTS) local loop is defined as the twisted pair of copper wires that most of us have coming into our home or business. This local loop is sometimes referred to as the *final 3 miles* or simply the *final mile*. The local loop has been *tuned* to our voice frequencies over the last 100 years and has a bandwidth of approximately 4000 hertz. This bandwidth includes two *guardbands* to prevent adjacent frequency interference. As can be seen in figure 3-2, the bandwidth available to the local loop circuit for actual voice analog transmission is about 3000 hertz.

The local loop wire pair consists of two wires and runs from a home or business to a local exchange carrier (LEC) central office (CO), which is also referred to as the *central exchange (CE)*. The CO provides voltage (–48 volts DC) for

FIGURE 3-1
Frequency range diagram.

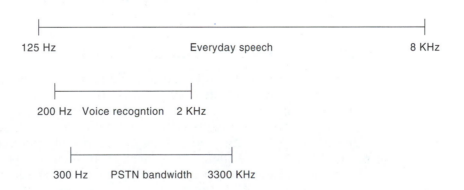

125 Hz   Everyday speech   8 KHz

200 Hz   Voice recogntion   2 KHz

300 Hz   PSTN bandwidth   3300 KHz

FIGURE 3-2 PSTN bandwidth.

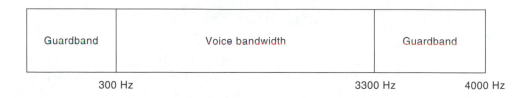

FIGURE 3-3 Local loop telephone circuit.

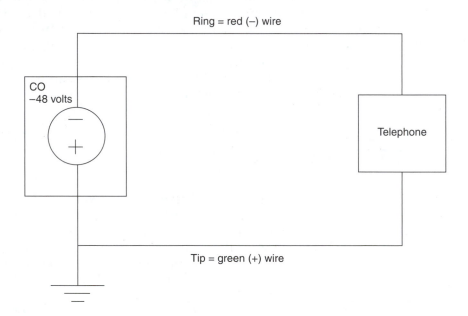

the telephones in our homes and businesses. The wires that make up a wire pair are identified as follows: The *tip* (green wire) is attached to the negative side of the CO 48-volt battery, and the *ring* (red wire) is attached to the positive side of the CO 48-volt battery. Figure 3-3 shows a basic local loop telephone circuit. In chapter 2, we learned that the CO provides the voltage for the telephone. We can now look a little more closely at the local loop in the form of a transmission line.

Transmission Lines and the Local Loop

The local telephone loop (also referred to as the *subscriber loop*) is the dedicated copper wire twisted pair connecting a telephone company central office (CO) in a locality to a customer home or business.

The loop resistance is critical in the local loop, and phone companies have had to *tune* the loop to transmit high-quality voice. Typically, companies have used 19 gauge (1.25-decibel attenuation per mile) to 26 gauge (3-decibel attenuation per mile) copper wire for the local loop. The average

FIGURE 3-4
Transmission line
model.

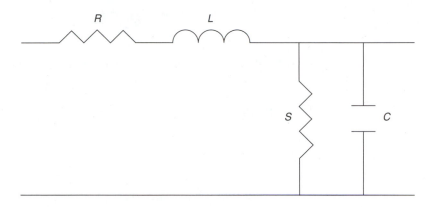

FIGURE 3-5 Two
wires separated by
insulation, forming a
capacitance.

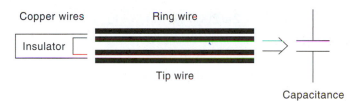

customer local loop is about 2 miles, and attenuation on this loop is ideally kept below 8 decibels (dB). We can look at a typical transmission line model and use it to represent a subscriber loop, as illustrated in figure 3-4.

Figure 3-4 shows that the inductance (L), resistances (R for series resistance and S for shunt resistance), and capacitance (C) are distributed throughout the model. These values can also cause signal loss and distortion. A local loop copper wire pair effectively forms a capacitance since two conductors (copper wire) are separated by an insulator (wire insulation). Shunt or mutual capacitive reactance is independent of wire gauge, and local loop wire pairs designed for voice have a capacitance value of about .083 microfarad/mile.

In addition to local loop cable, copper cables made for higher frequencies like those used for T-carrier systems are designed to provide a capacitance of .066 microfarad/mile. Figure 3-5 indicates how capacitance is formed when two conductors—in this case, wires—are placed close together separated by an insulator.

Capacitive reactance (X_C) can be calculated as:

$$X_C = \frac{1}{2\pi f C}$$

As frequency (f) increases, capacitive reactance drops. On long local loops (3 miles and greater), shunt capacitance values increase to the point where significant signal leakage occurs at frequencies greater than 1000 hertz. Looking at the formula, we realize that the higher the frequency the greater the leakage loss. Let us look at some examples.

●—**EXAMPLE 3.1**

●—**Problem**

A local loop is 1 mile long. Calculate the capacitive reactance for the loop at 2 kilohertz.

●—**Solution**   Using

$$X_C = \frac{1}{2\pi fC}$$

$$f = 2 \text{ kHz}$$

$$C = (1 \text{ mile})(.083 \text{ } \mu\text{F/mile}) = .083 \text{ } \mu\text{F}$$

$$X_C = \frac{1}{2\pi fC} = \frac{1}{2\pi(2000 \text{ Hz})(.083 \text{ } \mu\text{F})} = 959 \text{ } \Omega$$

●—**Problem**

This same local loop is extended to 3 miles. Calculate the new capacitive reactance for the loop at 2 kilohertz.

●—**Solution**   Using

$$X_C = \frac{1}{2\pi fC}$$

$$f = 2 \text{ kHz}$$

$$C = (3 \text{ miles})(.083 \text{ } \mu\text{F/mile}) = .249 \text{ } \mu\text{F}$$

$$X_C = \frac{1}{2\pi fC} = \frac{1}{2\pi(2000 \text{ Hz})(.249 \text{ } \mu\text{F})} = 319 \text{ } \Omega$$

In example 3.1, we can see that, by increasing the length of the loop by 2 miles, shunt capacitance drops by a factor close to 3.

In addition to length, higher frequencies also cause shunt capacitive reactance to increase. Now consider examples 3.2 and 3.3.

●—**EXAMPLE 3.2**

●—**Problem**

Let us increase the frequency in part b of example 3.1 to 3 kilohertz and calculate the capacitive reactance of the local loop.

●—**Solution** Using

$$X_C = \frac{1}{2\pi f C}$$

$$f = 3 \text{ KHz}$$

$$C = (3 \text{ miles})(.083 \text{ μF/mile}) = .249 \text{ μF}$$

$$X_C = \frac{1}{2\pi f C} = \frac{1}{2\pi(3000 \text{ Hz})(.249 \text{ μF})} = 213 \text{ Ω}$$

●—**EXAMPLE 3.3**

●—**Problem**

Let us now decrease the frequency to 1 kilohertz and calculate the capacitive reactance of the local loop.

●—**Solution** Using

$$X_C = \frac{1}{2\pi f C}$$

$$f = 1 \text{ KHz}$$

$$C = (3 \text{ miles})(.083 \text{ μF/mile}) = .249 \text{ μF}$$

$$X_C = \frac{1}{2\pi f C} = \frac{1}{2\pi(1000 \text{ Hz})(.249 \text{ μF})} = 639 \text{ Ω}$$

Now consider a voice conversation on the example 3.2 local loop. We know the frequency range of the local loop is approximately 300 hertz to 3300 hertz. We know that the human voice can produce frequencies of both 3 kilohertz and 1 kilohertz and that the average ear can hear these frequencies. At 1 kilohertz we have a shunt capacitive reactance of 639 ohms; and, at 3 kilohertz, we have a shunt capacitive reactance of 213 ohms. We see that more signal is lost due to capacitive shunting at the higher frequencies than at the lower frequencies. The listener will notice these differences; the lower

FIGURE 3-6 Loaded and unloaded loss.

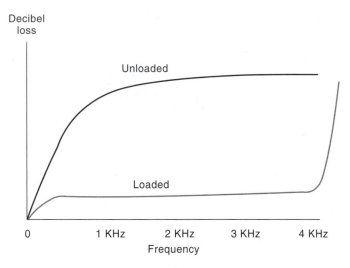

frequencies in a voice conversation will appear louder than the higher frequencies in a conversation.

Loading Coils

Both George Campbell at AT&T and Michael Pupin at Columbia University were working in 1899 on wire pair mutual capacitance problems. Both realized that, by adding a lump series inductance called a **loading coil,** resonance could be used to cancel the effects of shunt capacitive reactance and increase signal strength over long local loops. Michael Pupin ended up getting the patent; and, by late 1899, loading coils were being installed in the field on copper wire pairs longer than 3 miles.

Loading coils are a simple lump series inductance that produces an effect called *loading.* Loading increases the series inductance of the loop and effectively makes the loop a **low-pass filter,** increasing the impedance of the line, which drops signal attenuation. A typical 26-gauge local loop pair is loaded with a *26H88* loading coil every 6000 feet. The letter *H* designates a coil that is added every 6000 feet; *26* represents 26-guage wire; and *88* indicates that the inductance of the coil is 88 milliHenries (mH). This loading makes the loop perform as a low-pass filter and cuts the frequency off sharply at around 3.4 kilohertz. Loading coils work great for the low bandwidth requirements of voice but cause problems when we want to transmit data at higher bandwidths over these same wires. Figure 3-6 indicates the effect a loading coil will have on the frequency response of a local loop. At voice frequencies, the **cutoff frequency** (f_C) for a transmission line can be approximated as follows:

$$f_C \approx \frac{1}{\pi\sqrt{LDC}}$$

where

L = loading coil inductance

D = distance in miles between loading coils

C = capacitance per mile

Let us look at how to use this formula in example 3.4.

●—EXAMPLE 3.4

●—Problem

Calculate the cutoff frequency for a local 3-mile loop with 26H88 loading coils spaced every 6000 feet.

●—Solution   Using

$$f_C \approx \frac{1}{\pi\sqrt{LDC}}$$

L = 88 mH

D = 6000 feet ≈ 1 mile

C = .083 μF/mile

$$f_C \approx \frac{1}{\pi\sqrt{LDC}} = \frac{1}{\pi\sqrt{(88\ \text{mH})(1\ \text{mile})(.083\ \text{μF/mile})}} = 3724.5\ \text{Hz}$$

In example 3.4, we are well above the 3300-hertz cutoff we need for voice transmission and recognition so the loading coil has done its job. Notice, using this formula, that total loop distance is not used in the calculation—only distance between coils is required. In addition to H (6000 feet) load coil spacing, there are also B (3000 feet) spacing and D (4500 feet) spacing loading coils. By changing coil spacing along with coil inductance, the loop cutoff frequency can be adjusted or tuned to the proper value. In example 3.5, we can look at the effects of 22-millihenry B-spaced coils on a local loop.

●—EXAMPLE 3.5

●—Problem

22-millihenry loading coils are spaced every 3000 feet on a local loop. Calculate the cutoff frequency.

●─**Solution** Using

$$f_c \approx \frac{1}{\pi\sqrt{LDC}}$$

$L = 22$ mH

$D = 3000$ feet $\approx .5$ mile

$C = .083$ µF/mile

$$f_c \approx \frac{1}{\pi\sqrt{LDC}} = \frac{1}{\pi\sqrt{(22 \text{ mHz})(.5 \text{ mile})(.083 \text{ µF/mile})}} = 10534.5 \text{ Hz}$$

Notice in example 3.5, by reducing the distance between coils and decreasing the individual coil inductance values, that we can increase the cutoff frequency. There are three commonly used loading coils in the United States; the coil specifications are listed in table 3-1.

Loading coils have been used over the last 100 years and are an excellent way to tune a local loop to specific frequencies. As carriers move to provide high-bandwidth data services such as asymmetric digital subscriber line (ADSL) on the same local loop being used for voice, the low-pass filter characteristics of the loaded local loop provide significant bandwidth limitations. We can see that frequencies above 4000 hertz on loaded loops are blocked.

Bridged Taps

A **bridged tap** is an unterminated wire pair that sits in parallel to the main wire pair. Ideally, the local loop is a continuous wire point-to-point connection. At one time, the local loops were all set-up this way; but, with the growth of neighborhoods, new unused wire pairs got added. Typically, extra pairs are included though not initially used when cable is run down a street. When a new house is built, or a line is added, a phone company technician taps into one of the unused pairs. The technician typically does not cut the pair; the wires are just *tapped,* leaving the unterminated ends running down the street. This way, if the line is no longer needed, a technician can come out, remove the tap, and still use the pair for another customer farther down

TABLE 3-1
Commonly used loading coils.

Type	Inductance	Cutoff frequency (f_c)	Typical use
H88	88 mH	3.5–4 KHz	Local loops
H44	44 mH	5–5.6 KHz	Data circuits
B22	22 mH	10–11.2 KHz	Networks

FIGURE 3-7
Bridged-tap example.

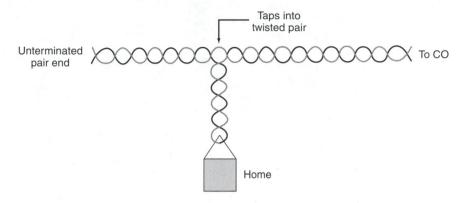

the street. This leaves a *bridged tap* with the *tap point* being where the technician spliced into the wire pair on the street, as illustrated in figure 3-7.

Bridged taps can create an impairment to the transmission system. A signal on the loop moves down the unterminated cable and will reflect back to the main pair, affecting the main signal. A bridged tap will typically not be noticed at voice transmission frequencies. All that is experienced is a slight increase in attenuation due to added capacitive load, which is usually so small that it is not detected by the human ear.

Analog Signals beyond the Local Loop

Analog transmission works fine for voice transmission on the local loop, and the existing copper pairs coming into our businesses should exist for another 10 to 20 years. Several companies have been working toward converting the copper twisted pair local loop to fiber or wireless. In addition, several cable television companies are now offering dialtone to their customers. Let us look at what happens to our analog voice transmission when it gets to the telephone company CO, as illustrated in figure 3-8.

FIGURE 3-8 Local loop and CO.

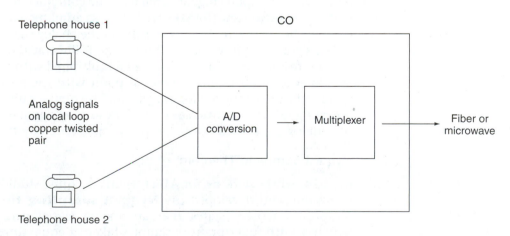

FIGURE 3-9
CODEC conversions.

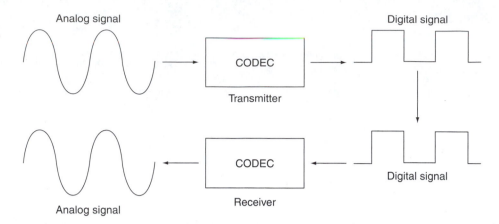

The analog voice signal is converted to a digital signal by something called a **CODEC**. A *CODEC,* short for *coder/decoder,* is a device used to convert analog signals to digital signals on one end and, on the other end, convert a digital signal back to an analog signal, as illustrated in figure 3-9.

Once the voice signal is digitized by the CODEC, it is then **multiplexed,** or combined, with other converted analog signals coming from other telephones or **analog modems** being served by the same CO. Once multiplexed, the calls are then sent out of the CO along their way on a higher-bandwidth transmission medium such as fiber or microwave.

Pulse Code Modulation (PCM) and Pulse Amplitude Modulation (PAM)

CODECs use a method called **pulse code modulation (PCM)** to convert the analog signals to digital bit streams. Pulse code modulation (PCM) is a multiplexing technique that uses *sampling* to obtain instantaneous voltage values at specific times in the analog signal cycle. Using PCM, the analog signal is combined with a sampling digital signal with the combined result being a **pulse amplitude modulated (PAM)** signal. A pulse amplitude modulated (PAM) signal is a snapshot of the instantaneous value of the analog signal being sampled at the digital signal sampling points. This process is illustrated in figure 3-10.

Figure 3-10 shows an analog signal multiplied with a digital pulse train instantaneous point by instantaneous point with the result being a PAM wave representation of the analog waveform. The digital pulse train determines the sampling rate; it is easy to see that, if the analog signal is not sampled enough, the analog signal will not be properly represented by the PAM signal.

Nyquist Sampling Theorem

In 1924, while working for AT&T, Henry Nyquist studied the PCM sampling technique and developed the **Nyquist sampling theorem.** The Nyquist sampling theorem states that an analog signal can be uniquely reconstructed, without error, from samples taken at equal time intervals if the sam-

FIGURE 3-10 PAM signal generation.

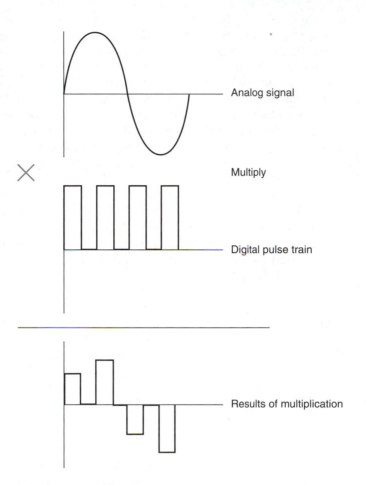

Analog signal

Multiply

Digital pulse train

Results of multiplication

pling rate is equal to, or greater than, twice the highest frequency component (f_{high}) in the analog signal or:

Sampling rate = $2(f_{high})$

Consider an example of Nyquist's theorem.

●—EXAMPLE 3.6

●—Problem
The public switched telephone network (PSTN) passes frequencies up to approximately 3300 hertz. Using the Nyquist sampling theorem, calculate the minimal sampling rate of the PSTN.

●—Solution
Sampling rate = $2(f_{high})$ = 2(3300 Hz) = 6600 samples per second

---

Using our calculation in example 3.6, we can see that, for minimal analog to digital conversion, an analog voice telephone line must be sampled minimally 6600 samples per second. Sampling at a rate of less than 6600 samples per second will not reproduce the signal properly. Sampling rates of greater than 6600 samples per second will produce more detail. Designers of the voice network used Nyquist's sampling theorem to determine the proper sampling rate. They knew they could not sample under the 6600 samples per second rate and also knew that going over the 6600 samples per second rate would produce higher quality. A PCM sampling rate of 8000 samples per second was selected.

Quantization is a technique that is used, along with a sampling rate, to generate a pulse code modulated (PCM) wave. Using quantization, the instantaneous voltage value of an analog signal is quantized into 2^8 (256) *discrete* signal levels. With each sample, the signal is instantaneously measured and adjusted to match one of the 256 discrete voltage levels. The technique is illustrated in figure 3-11.

Since quantization adjusts voltage levels to match one of 256 discrete voltage levels, it is easy to see that there will be some signal distortion. This signal distortion is known as **quantizing noise** and is greater for low-amplitude signals than for high-amplitude signals. A technique called **companding** is used to correct this problem. Companding is a method that compresses and divides the lower-amplitude signals into more voltage levels and provides more signal detail at the lower-voltage amplitudes. In North America and Japan, this is done using an algorithm called the μ-*law* (Greek letter "mhu"). Other countries in the world use the *A-law* algorithm so conversion is required when calls are made between countries using different algorithms.

Once a segment of an analog signal has been quantized and companded, it is then given an eight-bit binary code. This process is referred to as **encoding.**

After a single analog signal has been encoded, it is multiplexed, or combined, with 24 other encoded eight-bit signals. This generates a 192-bit (8×24) sequence for the 24 signals. A process called **framing** then adds one framing bit to create a 193-bit **frame.** The framing bits are used to keep the receiving device in *synchronization* with the frames it is receiving and to follow a 12-frame pattern that is repeated with each 12 frames. This sampling rate has determined the digital signal (DS) level system covered later in this chapter.

The Changing Local Loop

It is not economical at this time to run fiber into every home, but local-exchange carriers have been working to replace portions of the local loop with fiber by running fiber out from the CO into a **remote terminal (RT).**

FIGURE 3-11 PCM wave generation (note: not to scale).

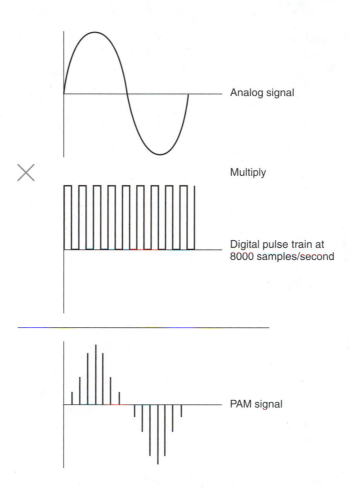

Analog signal

Multiply

Digital pulse train at 8000 samples/second

PAM signal

Remote terminals (RTs) are portable switching facilities outside of the CO and are typically pedestal boxes placed on a cement platform. One of the most common RTs in the field today is called a **multiple subscriber line carrier system** or **subscriber line carrier (SLC)-96** (*SLC* is pronounced *slick*). A multiple subscriber line carrier system, or SLC-96, takes 96, 64-kilobits-per-second analog voice or modem signals, converts them to digital, and then multiplexes them at the remote terminal. The remote terminal is connected to a central office using five, 1.544-megabits-per-second digital T-1 lines. This configuration is illustrated in figure 3-12.

In figure 3-12, it can be seen that four digital T-1 lines are used to carry the 96 digitized voice channels (1 T-1 line = 24 digitized voice channels so 4 T-1s are required to transmit 96 voice channels). The fifth T-1 line is used for protective switching and is a backup if one of the four fails.

FIGURE 3-12 SLC-96 field pedestal configuration.

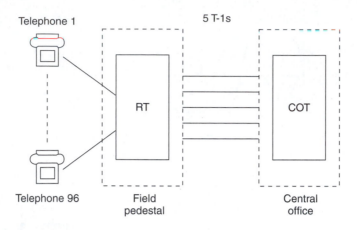

3.2 Multiplexing

Before the invention of the telephone, both Alexander Graham Bell and Thomas Edison were experimenting with ways to transmit more than one telegraph signal at a time over a single wire. They both realized that this was a critical piece if any communications network was to grow in the number of users. This is illustrated in figure 3-13. Three common ways to *multiplex* or combine multiple signals on the telephone network are analog or **frequency multiplexing**, **digital multiplexing**, and **wavelength division multiplexing (WDM)**.

Analog or Frequency Multiplexing

Analog or frequency multiplexing was used up until the early 1990s by long-distance carriers like AT&T and MCI and is still used today by the cable industry. The concept of *channel banks* was developed for analog multiplexing, and this concept is still used today for other types of multiplexing. Using *frequency multiplexing* to multiplex calls, each call was given a narrow range of

FIGURE 3-13 Multiplexing.

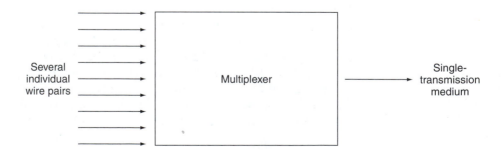

FIGURE 3-14 Single group formation.

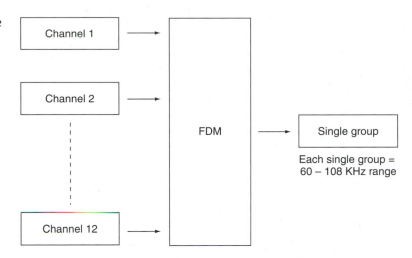

frequency in the available bandwidth. We know all voice call channels occupy the same frequency range—approximately 4000 hertz if we include individual call guardbands. If we want to combine a group of voice calls and separate them by frequency, we must translate the frequency of these individual call channels using a process called *single sideband, suppressed carrier (SSB-SC) modulation*. This technique allowed 10,800 individual voice call channels to be combined and transmitted over one coaxial cable copper pair. Let us look at how it was done.

Groups

Individual voice call channels are placed into **groups.** A group is a collection of 12 individual voice channels. If we have 12 channels per group and each channel uses 4000 hertz, we can calculate:

(12 voice channels) × (4000 Hz/voice channel) = 48,000 Hz

This 48 kilohertz is placed in the frequency range of 60 to 108 kilohertz, as illustrated in figure 3-14.

Supergroups

Individual groups are placed into **supergroups**. A supergroup is a collection of five groups, and each supergroup contains 60 individual voice channels. If we have five groups and each group uses 48 kilohertz, we can calculate:

(5 groups) × (48 KHz/group) = 240 KHz

This 240 kilohertz is placed in the frequency range of 312 to 552 kilohertz, as illustrated in figure 3-15.

FIGURE 3-15
Supergroup
formation.

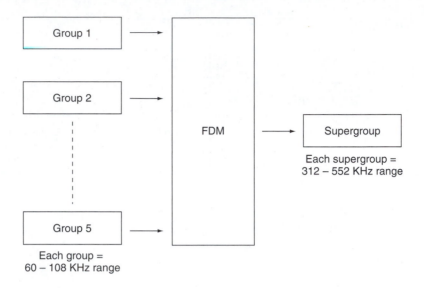

Each group =
60 – 108 KHz range

Each supergroup =
312 – 552 KHz range

FIGURE 3-16
Mastergroup
formation.

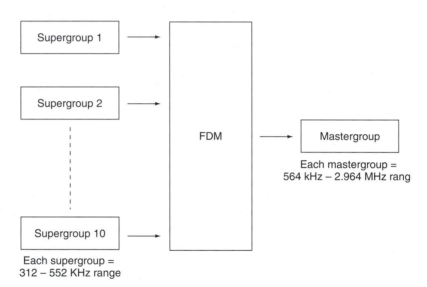

Each supergroup =
312 – 552 KHz range

Each mastergroup =
564 kHz – 2.964 MHz rang

Mastergroups

Individual supergroups are placed into **mastergroups.** A mastergroup is a collection of 10 supergroups and contains 600 individual voice channels. If we have 10 supergroups and each supergroup uses 240 kilohertz, we can calculate:

(10 supergroups) × (240 KHz/supergroup) = 2.4 MHz

This 2.40 megahertz is placed in the frequency range of 564 kHz to 2.964 megahertz, as illustrated in figure 3-16.

FIGURE 3-17
Jumbogroup
formation.

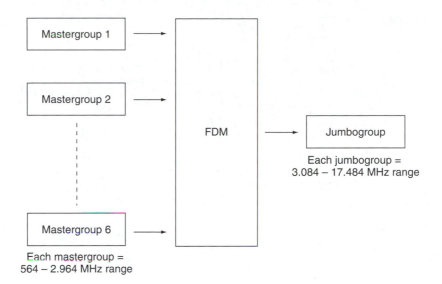

Jumbogroup

Each jumbogroup =
3.084 – 17.484 MHz range

Each mastergroup =
564 – 2.964 MHz range

Jumbogroups

Individual mastergroups are placed into **jumbogroups.** A jumbogroup is a collection of six mastergroups and contains 3600 individual voice channels. If we have six mastergroups and each mastergroup uses 2.4 megahertz, we can calculate:

(6 mastergroups) × (2.4 MHz/mastergroup) = 14.4 MHz

This 13.4 megahertz is placed in the frequency range of 3.084 to 17.484 megahertz, as illustrated in figure 3-17.

Jumbogroup Multiplex

The final multiplexing step involves combining individual jumbogroups, which are placed into **jumbogroup multiplexes.** A jumbogroup multiplex is a collection of three jumbogroups and contains 10,800 individual voice channels.

Frequency division multiplexing is now considered obsolete technology on the voice-based telecommunications network. Analog signals are more sensitive to noise and other signals that can cause problems along the transmission path. They have been replaced with digital multiplexers.

Digital Multiplexing

Digital signals are combined or multiplexed typically using one of two techniques—**time division multiplexing (TDM)** and **statistical time division multiplexing (STDM).**

FIGURE 3-18 TDM framing.

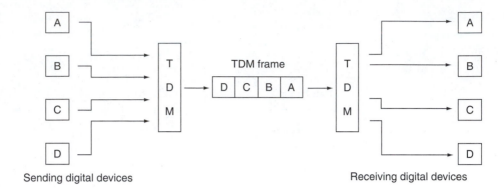

FIGURE 3-19 TDM framing showing wasted slots.

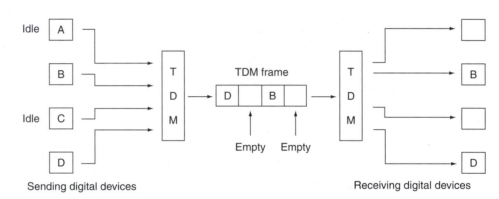

Time Division Multiplexing

Time division multiplexing (TDM) is a digital signal multiplexing technique that allows multiple devices to communicate over the same circuit by assigning time slots for each device on the line. Devices communicating using TDM are typically placed in groups that are multiples of three.

Each device is assigned a time slot where the TDM will accept an eight-bit character from the device. A TDM frame is then built and transmitted over the circuit. Another TDM on the other end of the circuit de-multiplexes the frame, as illustrated in figure 3-18.

Statistical Time Division Multiplexing (STDM)

TDM tends to waste time slots because a time slot is allocated for each device regardless of whether that device has anything to send. For example, in a TDM system, if only two of four devices want to send and use frame space, the other two devices will not have anything to send. This is illustrated in figure 3-19.

Idle devices do not require frame space, but their time slots are still allocated and will be transmitted as empty frames. This is not an efficient use of bandwidth. A statistical time division multiplexer (STDM or STATDM or

FIGURE 3-20 STDM multiplexing.

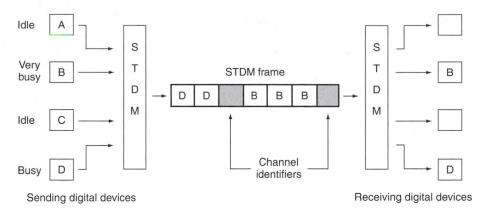

STAT MUX) does not assign specific time slots for each device by using a process called *statistical time division multiplexing.* Statistical time division multiplexing is a digital multiplexing technique that adds an address field to each time slot in the frame and does not transmit empty frames. Only devices that require time slots get them.

Statistical time division multiplexing uses *dynamic* time-slot lengths that are variable. Communicating devices that are very active will be assigned greater time-slot lengths than devices that are less active. If a device is idle, it will not receive any time slots. For periods of much activity, STDMs have buffer memory for temporary data storage. STDM is illustrated in figure 3-20.

Each STDM transmission carries *channel identifier* information. Channel identifier information includes the source device address and a count of the number of data characters that belong to the listed source address. Channel identifiers are extra and considered **overhead.** In a transmission system, overhead is transmitted information that is not actual desired content (voice, data, video, etc.). To reduce the cost of channel identifier overhead, it makes sense to group large numbers of characters for each channel together.

Wavelength Division Multiplexing (WDM)

As bandwidth requirements continue to grow for both the legacy PSTN and the emerging Internet, most of the high-bandwidth backbone transmission is being done with fiber optics and a method called *wavelength division multiplexing (WDM).* WDM functions very similarly to **frequency division multiplexing (FDM).** With FDM, different frequencies represent different communications channels with transmission done on copper or microwaves. Wavelength division multiplexing is a multiplexing technique that uses **wavelength** instead of frequency to differentiate the different communications channels.

FIGURE 3-21
Wavelength
measurement.

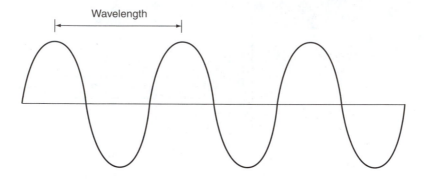

FIGURE 3-22
Wavelength division
multiplexing.

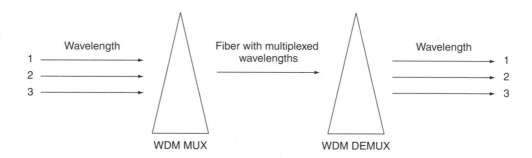

Wavelength

Analog waves, from voice to radio to light to X rays, are sinusoidal in nature; and *wavelength,* represented by the Greek letter lambda (λ) is a distance measurement usually expressed in meters. Wavelength is defined as the distance in meters of one sinusoidal cycle, as illustrated in figure 3-21.

Wavelength indicates the color of light. For example, the human eye can see light ranging in frequency from approximately 380 nanometers (dark violet) to approximately 765 nanometers (red). WDM multiplexers use wavelength, or color, of light to combine signal channels onto a single piece of optical fiber. Each WDM signal is separated by wavelength *guardbands* to protect from signal crossover. One of WDM's biggest advantages is that it allows incoming high-bandwidth signal carriers that have already been multiplexed to be multiplexed together again and transmitted long distances over one piece of fiber, as illustrated in figure 3-22.

In addition to WDM, systems engineers have developed even higher capacity **dense wavelength division multiplexing (DWDM)** systems. Dense wavelength division multiplexing (DWDM) is an even higher capacity multiplexing technique that also uses wavelength to combine signal channels onto a single piece of optical fiber.

In early 2000, a Siemens' prototype system transmitted 80 gigabits per second of information on 40 individual wavelengths in a DWDM system.

This system can transmit 3.2 terabits per second of information over 400 kilometers on a single strand of optical fiber. As backbone bandwidth requirements continue to grow, these WDM and DWDM systems will significantly reduce long-haul bandwidth bottlenecks.

3.3   The Digital Signal (DS) Level System

We know that, to convert an analog signal to a digital signal, the analog signal is sampled 8000 times per second and, after matching the instantaneous voltage sample level to one of 256 discrete levels, an eight-bit code is generated for each sample. If we multiply the sample rate by the bit code, we get:

(8000 samples/second)(8 bits/sample) = 64,000 **bits per second (bps)**

So, we can say that a single analog voice channel, after conversion from analog to digital, requires 64 kilobits per second of digital bandwidth. This 64 kilobits per second is referred to as *digital* signal level 0 or **DS-0.** DS-0 is the 64 kilobits per second basic building block or channel for the existing digitally multiplexed T-carrier system in the United States and the digital E-carrier system used in Europe.

Voice calls are digitally multiplexed using either time division multiplexing or statistical time division multiplexing. Calls are grouped in a way similar to frequency division multiplexing. Let us look at how this is done.

Digroups or DS-1 Signals

Individual analog voice call channels converted to digital require a bit rate of 64 kilobits per second each. Twenty-four 64-kilobits-per-second digital voice channels are multiplexed into digroups *or* **DS-1** signals. A DS-1 signal is a combination of 24 DS-0 channels with added overhead. If we have 24 DS-0 signals per DS-1 signal and each channel is 64 kilobits per second, we can calculate:

(24 DS-0 signals) × (64 Kbps/DS-0 signal) = 1.536 Mbps

Adding overhead consisting of timing and synchronization bits brings the DS-1 bit rate to 1.544 megabits per second, as illustrated in figure 3-23.

DS-1 Overhead

We have described the process of *encoding* where an analog signal is sampled 8000 times per second, quantized into one of 256 discrete signal levels, companded, then given an eight-bit binary code. After a single analog signal sample has been encoded, it is multiplexed with 24 other encoded eight-bit sample signals. This generates a 192-bit (8 bits/sample signal × 24 sample signals) sequence for the 24 sample signals. A process called *framing* then adds one framing bit to create a 193-bit frame, as illustrated in figure 3-24.

FIGURE 3-23 DS-1 formation.

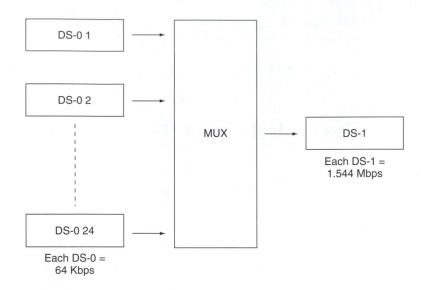

FIGURE 3-24 DS-1 with overhead.

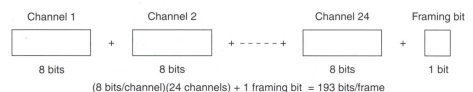

FIGURE 3-25 Masterframe.

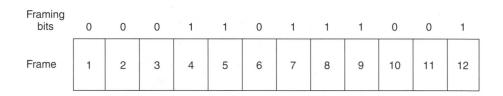

As can be seen in figure 3-24, a *frame* is just a combination of 24 sample signals with a single framing bit added and is 193 bits in length. The framing bits are used to keep the receiving device in synch with the frames it is receiving. A **masterframe,** also called a **superframe,** is a grouping of 12 frames in the DS-carrier system. Included within each masterframe is a 12-bit frame pattern from the 12 grouped 193-bit frames. This 12-bit frame pattern carries a bit pattern of *000110111001* and repeats itself with each masterframe. The bit pattern is illustrated in figure 3-25. This masterframe bit pattern is used for *synchronization*.

Remember that each channel is sampled 8000 times per second, so a single frame represents one eight-thousandth of 24 individual channels or telephone calls. We can also say that, in one second, a DS-1 signal transmits

FIGURE 3-26 DS-2 formation.

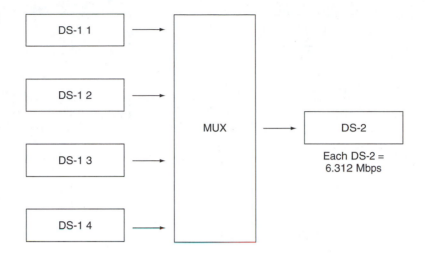

8000 193-bit frames. We can use these numbers to calculate the true DS-1 bit rate, which includes both data and overhead (framing) bits:

$$\text{DS-1 bit rate} = (8000 \text{ frames/second})(193 \text{ bits/frame})$$
$$= 1.544 \times 10^6 \text{ bps} = 1.544 \text{ Mbps}$$

DS-2 Signal

Each DS-1 signal carries a bit rate of 1.544 megabits per second. A **DS-2** signal is four 1.544-megabits-per-second digital DS-1 signals multiplexed together into one signal. If we have four DS-1 signals per DS-2 signal and each DS-1 signal is 1.544 megabits per second, we can calculate:

$$(4 \text{ DS-1 signals}) \times (1.544 \text{ Mbps/DS-1 signal}) = 6.176 \text{ Mbps}$$

Adding overhead consisting of timing and synchronization bits brings the DS-2 bit rate to 6.312 megabits per second, as illustrated in figure 3-26.

DS-3 Signal

Each DS-2 signal carries a bit rate of 6.312 megabits per second. A **DS-3** signal is seven 6.312-megabits-per-second digital DS-2 signals multiplexed together into one signal. If we have seven DS-2 signals per DS-3 signal and each DS-2 signal is 6.312 megabits per second, we can calculate:

$$(7 \text{ DS-2 signals}) \times (6.312 \text{ Mbps/DS-2 signal}) = 44.184 \text{ Mbps}$$

Adding overhead consisting of timing and synchronization bits brings the DS-3 bit rate to 44.736 megabits per second, as illustrated in figure 3-27.

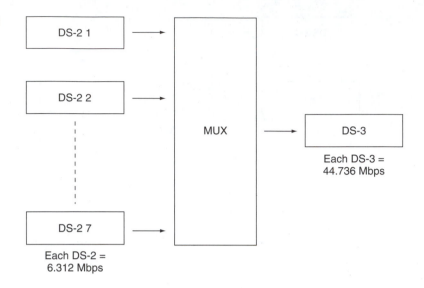

FIGURE 3-27 DS-3 formation.

DS-4 Signal

Each DS-3 signal carries a bit rate of 44.736 megabits per second. A **DS-4** signal is six 44.736-megabits-per-second digital DS-3 signals multiplexed into one signal. If we have six DS-3 signals per DS-4 signal and each DS-3 signal is 44.736 megabits per second, we can calculate:

(6 DS-3 signals) × (44.736 Mbps/DS-3 signal) = 268.416 Mbps

Adding overhead consisting of timing and synchronization bits brings the DS-4 bit rate to 274.176 megabits per second, as illustrated in figure 3-28.

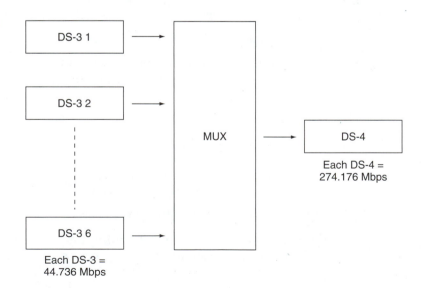

FIGURE 3-28 DS-4 formation.

FIGURE 3-29 DS-5 formation.

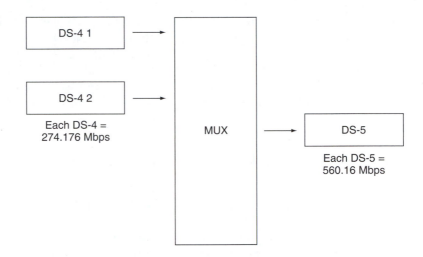

DS-5 Signal

Each DS-4 signal carries a bit rate of 274.176 megabits per second. A **DS-5** signal is two 274.176-megabits-per-second digital DS-4 signals multiplexed into one signal. If we have two DS-4 signals per DS-5 signal and each DS-4 signal is 274.176 megabits per second, we can calculate:

(2 DS-4 signals) × (274.176 Mbps/DS-4 signal) = 548.352 Mbps

Adding overhead consisting of timing and synchronization bits brings the DS-5 bit rate to 560.16 megabits per second, as illustrated in figure 3-29. One DS-5 channel can carry 8064 voice channels. Table 3-2 shows the DS data rates and how they correspond to the North American T-carrier system.

TABLE 3-2 How DS data rates correspond to the North American T-carrier system.

DS #	# DS-0 signals	bps	T-carrier equiv.
DS-0	1	64 Kbps	—
DS-1	24	1.544 Mbps	T-1
DS-1C	48	3.152 Mbps	T-1C
DS-2	96	6.312 Mbps	T-2
DS-3	672	43.736 Mbps	T-3
DS-4	4032	273.176 Mbps	T-4
DS-5	8064	560.16 Mbps	T-5

Looking at table 3-2, it is easy to see that the DS-0 signal level is the foundation for the entire T-carrier hierarchy in North America. Notice one DS-1 line is the equivalent of 24 DS-0 64-kilobits-per-second DS-0 voice channels. Also notice that one DS-2 line is the equivalent of four DS-1 lines or 96 DS-0 voice channels. Copper wire pairs can be used to transmit at levels up to DS-2. At levels above DS-2, coaxial cable, fiber, or microwaves are used.

3.4 European E-Carrier Hierarchy

The European or **E-carrier** system is a digital transmission technique that is slightly different than the North American T-carrier system format. With the E-carrier system, we are still taking individual voice call analog signals and converting them to a digital signal by sampling the analog signal 8000 times per second and, after matching the instantaneous voltage sample level to one of 256 discrete levels, generating an eight-bit code for each sample. We are also still dealing with the fundamental DS-0 building block of 64 kilobits per second of digital bandwidth per single analog voice channel that we used for the T-carrier system. The differences between E-carrier and T-carrier deal with the number of channels and how these channels are used. Consider a European E-1 system and how it compares to a North American T-1 system.

The E-carrier system starts by multiplexing 32 DS-0 channels together to form an E-1 circuit, while the North American T-carrier system multiplexes 24 DS-0 channels to form a T-1 circuit. The 32 DS-0 channels of an E-1 circuit combine from channel 0 up to channel 31. Channel 0 is used for framing (synchronization), channels 1 through 15 and 17 through 31 are used for individual DS-0 channels, and channel 16 is reserved and not used. This is shown in table 3-3.

This system is also referred to as the *30 plus 2 system* because an E-1 signal consists of 30 DS-0 signals used for voice plus channel 0, which is used for overhead, and channel 16, which is not used at all. In the European system, all synchronization (framing) is handled by channel 0 so framing bits are not required on individual DS-0 channels.

TABLE 3-3 E-1 channel uses.

E-1 channel	Use
Channel 0	Framing (synchronization)
Channels 1–15	DS-0 signals
Channel 16	Reserved
Channels 17–31	DS-0 signals

TABLE 3-4 How DS data rates correspond to the European E-carrier system.

# DS-0 signals	bps	Multiplexed	E-carrier designation
1	64 Kbps	1 DS-0	—
32	2.048 Mbps	32 DS-0	E-1
128	8.448 Mbps	4 E-1s	E-2
512	33.368 Mbps	16 E-1s	E-3
2048	139.264 Mbps	4 E-3s	E-4
8192	565.148 Mbps	4 E-4s	E-5

We can calculate the signal rate for an E-1 circuit as follows:

$$\text{E-1 carrier bit rate} = (64 \text{ Kbps/channel})(32 \text{ channels}) = 2.048 \times 10^6 \text{ bps}$$
$$= 2.048 \text{ Mbps}$$

E-2 through E-5 are carriers in increasing multiples of the E-1 format. Table 3-4 shows the E-carrier DS data rates and how they correspond to the European E-carrier system.

3.5 Synchronous Optical Network (SONET)

In the United States, T-carriers are rapidly being replaced with **synchronous optical network (SONET)** systems. Synchronous optical network (SONET) transmission is a technique that transmits at rates up to a little less than 10 gigabits per second (Gbps). Internationally, the SONET equivalent is called *synchronous digital hierarchy (SDH)*. Both SONET and SDH systems consist of rings of fiber capable of carrying very high bit rates over long distances. Copper has been replaced by fiber to interconnect most COs in the United States at bit rates ranging from the SONET base rate of 51.84 megabits per second up to 9.95328 gigabits per second. The base SONET standard bit rate is 51.84 megabits per second and is referred to as *optical carrier (OC)-1* or *synchronous transport level (STS)-1*. Both **optical carrier (OC)** and **synchronous transport signal level (STS)** are terms used interchangeably to indicate SONET bit rate levels.

SONET uses a *synchronous* structure for framing that allows multiplexed pieces down to individual DS-0 channels to be pulled off a SONET signal without having to de-multiplex the entire SONET signal. Table 3-5 shows SONET bit rates.

The OC-1 base is used for all higher level SONET specifications. For example, a SONET specification of OC-48 can be calculated by taking the

TABLE 3-5 SONET
bit rates.

SONET spec	bps rate	# DS-0 channels
OC-1	51.84 Mbps	672
OC-48	2.48832 Gbps	32,256
OC-192	9.95328 Gbps	129,024

OC-1 base rate of 672 DS-0 channels and multiplying it by the OC-48 suffix of 48:

OC-48 DS-0 equivalent = (48)(672 DS-0 channels)
= 32,256 DS-0 channels

We can do the same calculation for the OC-192 specification:

OC-192 DS-0 equivalent = (192)(672 DS-0 channels)
= 129,024 DS-0 channels

It is common to run SONET rings CO to CO with all SONET-connected COs having SONET multiplexers that can de-multiplex all the way down to an individual DS-0 channel level without having to de-multiplex the entire SONET frame.

3.6 Data Transmisson on T-1 Carriers

The DS-1 or T-1 is the first multiplexed unit in the DS hierarchy; and, by looking at it in more detail, we can get a better idea of how the entire system works. The T-1 carrier uses time division multiplexing and was designed for voice call transmission. When using it for data, we would think it could be used to achieve a data bit rate of 64 kilobits per second over a T-1 carrier. Looking a little closer, we see that data on T-1 carriers is transmitted in the form of only seven-bit words; all eight bits are not used. Why?

Remember that the T-carrier system was initially designed for voice. The first signal synchronization used for the T-1 carrier substituted a single in-band signaling bit, used for control, for each of the 24 channels in every sixth frame. This means that, in the sixth and twelfth frames of every T-1 carrier masterframe, there is a bit used for in-band signaling.

Borrowing this single bit between the sixth and twelfth frames of a T-1 carrier masterframe and using it for in-band signaling is referred to as **bit robbing.** Bit robbing is usually not a problem when transmitting voice. Even though the signal is slightly distorted, the listener on the receiving end cannot perceive the distortion. However, this is a major problem when transmitting data as any data received with missing bits will be distorted and re-

FIGURE 3-30
Sample T-1 pulse
train.

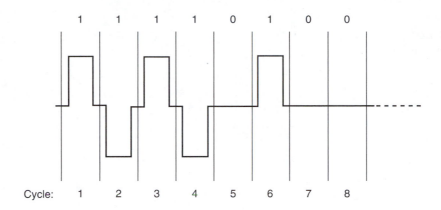

ceived incorrectly. To eliminate the problem caused by bit robbing, data on the T-1 carrier is limited to seven bits per frame in all frames. By decreasing the number of bits transmitted, the data bit rate is reduced.

T-1 data channel bit rate = (8000 frames/second)(7 bits/channel) = 56 Kbps

56-kilobits-per-second clear channel capability is the term used to refer to the T-1 carrier single channel maximum data bit rate.

T-1 Carrier Pulse Cycles

Looking more closely at a T-1 carrier signal, we see that there are negative and positive pulses combined in the digital pulse train. A sample T-1 signal pulse train is shown in figure 3-30.

Alternating positive/negative pulse trains (bipolar) has been shown to produce fewer transmission errors than using all positive or all negative pulse trains. These pulses are used to represent binary *1*s, and each pulse, when nonzero, is positive half the nonzero cycle (50 percent) and negative half the nonzero cycle. We can look at an example of a positive (cycle 1) and negative (cycle 4) pulse from figure 3-30 in figure 3-31. In figure 3-31, T represents the *period,* or time it takes to complete a single pulse cycle. We can calculate the *percent duty cycle* using the following equation:

$$Percent\ duty\ cycle = \frac{time\ in\ cycle\ signal\ not\ zero}{cycle\ period} \times 100$$

The pulses here are not zero for one half of the pulse period and have a 50 percent duty cycle.

Let us go back now and look at the original pulse train diagram and look at each cycle, as shown in table 3-6. We can now see that, if a pulse is present within a cycle time slot, whether positive or negative, it represents a *1* bit, and, if no pulse is present, it represents a *0* bit.

FIGURE 3-31
Sample T-1 positive-
and negative-going
pulses.

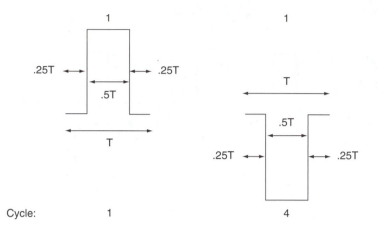

TABLE 3-6 T-1
pulse train cycles.

Cycle	Cycle pulse detail
1	Positive pulse representing a binary *1*.
2	Negative pulse representing a binary *1*; notice pulse polarity inversion when compared to previous *1* value pulse *1*.
3	Positive pulse representing a binary *1*; notice pulse polarity inversion when compared to previous *1* value pulse *2*.
4	Negative pulse representing a binary *1*; notice pulse polarity inversion when compared to previous *1* value pulse *3*.
5	No pulse present representing a binary *0*.
6	Positive pulse representing a binary *1*; notice pulse polarity inversion when compared to previous *1* value pulse *4*.
7	No pulse present representing a binary *0*.
8	No pulse present representing a binary *0*.

Bipolar with 8-Zero Substitution (B8ZS)

T-1 lines that are not constantly active (having binary *1*s) will have timing problems because actual pulses are also used for signal synchronization by the receiver. To add synchronization on "quiet" T-1 lines, a technique called **bipolar with 8-zero substitution (B8ZS)** has been developed. Bipolar with 8-zero substitution (B8ZS) is a T-1 technique that adds pulses by substituting eight zero-bit groups with one of two specific eight-bit binary codes.

FIGURE 3-32 B8ZS substitution with most previous *1* pulse a positive-going pulse.

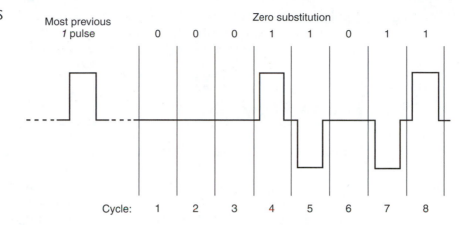

B8ZS Substitution with Most Previous *1* Pulse a Positive Going Pulse

When the transmitter gets a string of eight zeros and the most previous *1* pulse was a positive-going pulse, the eight-bit pulse sequence is substituted for the eight-zero sequence, as illustrated in figure 3-32. Notice in the figure there is a polarity discrepancy in this substituted pulse sequence. Pulses 5 and 7 are sequential *1* pulses and are both negative going—they do not alter in polarity.

B8ZS Substitution with Most Previous *1* Pulse a Negative Going Pulse

When the transmitter gets a string of eight zeros and the most previous *1* pulse was a negative-going pulse, the eight-bit pulse sequence is substituted for the eight-zero sequence, as illustrated in figure 3-33.

Notice that there is also a bipolar polarity discrepancy in the substituted pulse sequence illustrated in figure 3-33. Again, pulses 5 and 7 are sequential

FIGURE 3-33 B8ZS Substitution with most previous *1* pulse a negative-going pulse.

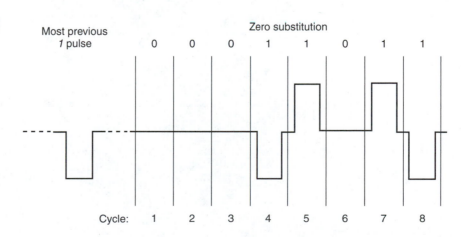

1 pulses. In this case, they are both positive going and do not alter in polarity. T-1 receivers are designed to detect both of these bipolar polarity discrepancies and substitute strings of eight zeros whenever one is detected.

3.7 Frame Relay

Frame relay is a packet-switching networking technology designed to transfer traffic between data networks. **Packet-switched networks** are networks designed not to use dedicated circuit line connections like the PSTN. In a packet-switched network, the information (which could be a data file, video, or voice) is separated into packets; and the packets are what is delivered across the network. Packets can be allowed to move individually through a frame-relay network and find their own way or can be sent through the network following the same path. Because of multiple-path redundancy, packet-switched networks like frame relay and the earlier X.25 network are often represented as a cloud. This is shown in figure 3-34.

Data traffic in general is commonly referred to as *bursty.* Consider how you surf the Internet, you call up a Web site, there is a burst of traffic downstream to your machine, and then you may spend a few minutes looking at the site before clicking another link and sending another burst of traffic down to your machine. Frame-relay circuits are designed for bursty networks. There are no standard bit rate offerings for frame relay; they vary from telecommunications provider to provider.

FIGURE 3-34
Frame-relay cloud.

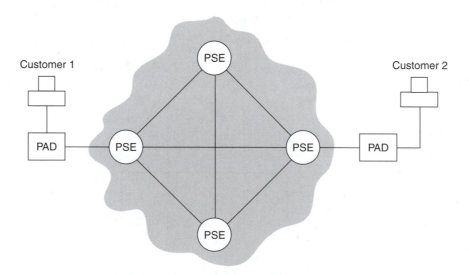

Frame Relay Transmission

Let us look at how data moves through the network from customer 1 to customer 2 in figure 3-34.

Step 1. Customer 1 sends data to the **packet assembler/disassembler (PAD).**

Step 2. The *packet assembler/disassembler (PAD)* is a computer-running software that will take the data from customer 1 and break it down into packets, or pieces of data, called *frames.* A PAD does not have to be dedicated to a single user; many users can be attached to a PAD with the PAD multiplexing frames together from multiple users.

Step 3. The PAD routes the created frames into the frame-relay cloud by passing the frames off to a node called a **packet-switching exchange (PSE).** The PSE is a computer, also commonly referred to as a **node,** that routes the frames though the network toward the destination by passing the frames from PSE to PSE within the frame cloud.

Step 4. The frame is routed through the attached PSEs within the frame cloud and arrives at the PAD attached to customer 2.

Step 5. The PAD attached to customer 2 disassembles the transmitted frame or frames and rebuilds the data, passing it off to customer 2.

In frame-relay networks, the connections formed within the network cloud are called **permanent virtual circuits (PVCs).** PVCs are called *virtual* because there is no direct circuit built through the network cloud; there is only a virtual—sometimes referred to as a *logical*—connection from one point to another across the network. Connections are shared with other frame-relay users at a specified bit rate called the **committed information rate (CIR).** CIR is the guaranteed throughput bit rate for a frame-relay PVC.

Frame Structure

Frame relay is considered to be a **high-level data link control (HDLC),** protocol-based, network-transmission technology. HDLC is a set of protocols or rules for transmitting data between points on a network of connected points commonly called nodes. Using HDLC, the PAD organizes data into frames with frame structure, as illustrated in figure 3-35. In the figure, notice that the frame does not just include data; the data is encapsulated by extra address and control information. This extra address and control information added to the data within the frame is called *overhead.* The frame header contains a data link connection identifier (DLCI) that the PSEs use to route each frame along the virtual path that has been set up. The data section of the

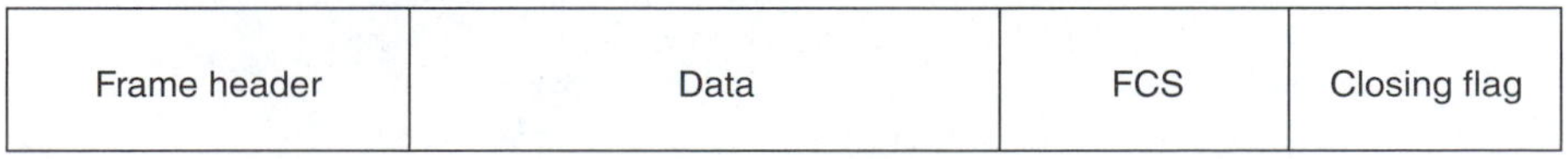

FIGURE 3-35
Frame-relay frame structure.

| Frame header | Data | FCS | Closing flag |

frame is variable in length up to 1610 bytes; the (FCS) is used to ensure data delivered is correct, and the closing flag indicates the frame has ended.

In contrast to earlier packet-switched networks like X.25, frame relay relies on high-quality transmission through the cloud and depends on the customer equipment to check frames for errors occur when they are received. Today's modern networks are efficient enough that not many transmission errors occur and the error correction can be handled easily by the end devices. By handling the transfer this way, throughput from customer device to device is very fast.

3.8 Asynchronous Transfer Mode

Asynchronous transfer mode (ATM) is a packet-switched, network-transmission technique that is used to connect data, voice, and video networks over long distances. ATM standards development started in 1984 with involvement by the Telecommunication Standardization Section of the International Telecommunications Union (ITU-T), the ATM Forum, and the **American National Standards Institute (ANSI).** ATM is extremely flexible and will work over almost any transmission medium at different bit rates. The technology is widely used over fiber by telephone companies in the United States and has been designed to be carried over SONET. ATM has become a popular standard for local, campus, and wide area networks because it allows one technology to be used for all levels of a network. ATM networks are also modeled as a cloud with redundant connects through the cloudlike frame relay, as illustrated in figure 3-36. ATM is connection-oriented, and a virtual path is set up for each transmission through the network, which makes it a good transmission medium not just for data but also for time-sensitive voice and video.

FIGURE 3-36 ATM cloud.

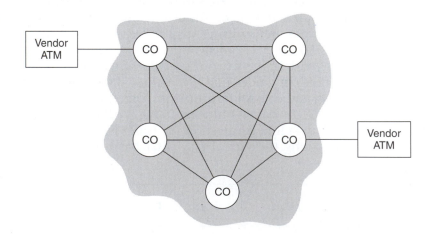

ATM uses asynchronous time division multiplexing to transmit fixed-length 53-byte packets called **cells** at popular common speeds of 155.52 megabits per second or 622.08 megabits per second from node to node within the ATM cloud. In the previous section, we saw that frame relay uses variable-length packets called frames. Processing variable-length frames can cause some slowdown because the equipment that is used to do the processing must determine where each frame starts and stops. By fixing the ATM cell size, processing equipment does not have to look specifically for the start and end of an ATM cell; the equipment knows each cell is 53 bytes long.

Cell Structure

The ATM cell has been made small at 53 bytes to allow for high-speed packet switching as the cells move through the network. ATM is considered connection-oriented; and, once a connection has been established, all cells will travel the same path through the network. The ATM cell structure is shown in figure 3-37.

The **ATM cell header** is five bytes long and identifies the cells belonging to a virtual channel by including a virtual circuit address. The virtual circuit address allows cells to be routed through the network on the correct virtual channel. The cell header also contains an error-correction byte that is used to detect and correct header errors. The header is made small for performance reasons and only contains the minimal amount of information to get the cell to its proper destination on the proper path. Header and trailer information will be different depending on the type of data being transmitted.

The **payload** section of an ATM cell is the actual data that is being transmitted. Payload for a single ATM cell will always be 48 bytes and is moved through the network with no devices along the transmission path performing any type of processing on the payload section. The selection of the 48-byte payload size was not selected without controversy. As the international standard was being developed, it was determined that a payload size between 32 and 64 bytes would work well with existing network components. The Europeans wanted to use a 32-byte payload, while the United States and Japan wanted to use 64 bytes. The standards groups ended up selecting a 48-byte payload size as a compromise.

Due to the reliability and low error rates on ATM transmission systems, the ATM network cloud provides no error detection, error correction, or any kind of retransmission services. All payload checking and correction are done by the end devices to save time and to support the high bandwidths associated with ATM.

FIGURE 3-37 ATM cell structure.

Header	Payload
5 bytes	48 bytes

Summary

1. The *PSTN* has been tuned to a frequency bandwidth of 300 to 3300 hertz.

2. The local loop is also referred to as the *final mile*.

3. *Capacitive shunting* causes more signal loss at higher frequencies than at lower frequencies.

4. *Loading coils* are added to the local loop to cancel the effects of capactive shunting.

5. The *H88* (88 millihertz) is the most common local loop loading coil added every 6000 feet.

6. *Bridged taps* are unterminated wire pairs that sit in parallel to main wire pairs.

7. *CODECs* are used to convert analog signals to digital and digital signals to analog using *pulse code modulation (PCM)*.

8. The *Nyquist sampling theorem* states that an analog signal must be sampled at at least twice its highest frequency for sample reproduction.

9. *Quantization* is used, along with a sampling rate, to generate a PCM wave.

10. *Multiplexers* are used to combine multiple signals on the telephone network.

11. A *SLC-96* is used to combine and multiplex 96, 64-kilobits-per-second analog voice lines.

12. *Analog* or *frequency multiplexing* was used until the early 1990s by long-distance carriers in the United States.

13. *Digital multiplexing* is typically done by *time division* or *statistical time division*.

14. *Wavelength division multiplexing (WDM)* uses light wavelength to combine multiple communications channels.

15. The *digital signal level 0 (DS-0)* 64-kilobits-per-second digital bandwidth is the basic building block for the existing digital *T-carrier* system used in the United States and for the existing digital *E-carrier* system used in Europe.

16. T-carrier systems in the United States are rapidly being replaced by *synchronous optical carrier (SONET)* systems.

17. The base SONET bit rate is 51.84 megabits per second and is referred to as *optical carrier 1* (OC-1) or *synchronous transport signal level 1* (STS-1).

18. T-1 carriers use *bit robbing* for in-band signaling; and, as a result, data on T-1 carriers is limited to seven bits per frame.

19. T-1 carrier pulses *alternate* from positive to negative to produce fewer transmission errors.

20. T-1 carriers use *B8ZS* to maintain timing on less-active lines.

21. *Frame relay* is a packet-switched networking technology designed to transfer traffic between data networks.

22. Frame relay uses *packet assemblers/disassemblers (PADs)* and *packet-switching exchanges (PSEs)* to move frames through the frame-relay cloud.

23. Frame-relay networks connected within the network cloud are called *permanent virtual circuits (PVCs)*.

24. Frame relay relies on *error correction* occurring on the end devices outside of the cloud.

25. *Asynchronous transfer mode (ATM)* uses asynchronous time division multiplexing to transmit fixed-length 53-byte cells through an ATM network cloud.

26. ATM is considered a *connection-oriented* transmission method.

27. The ATM *header* is five bytes long and contains minimal information to route a cell through the network.

28. The ATM *payload* is actual data that is being transmitted and is fixed at 48 bytes.

Review Questions

Section 3.1

1. Typical human voice frequency is centered around _____ hertz.
   a. 200–400
   b. 400–600
   c. 1400–1800
   d. 2800–3000

2. The PSTN local loop has been tuned to the _____ hertz frequency range.
   a. 125-Hz–8-KHz
   b. 300-Hz–3.3-KHz
   c. 200-Hz–2-KHz
   d. 400-Hz–800-KHz

3. _____ have been added to tune the local loop to the proper voice frequency range.
   a. Shunt capacitors
   b. Bridged taps
   c. Unterminated wire pairs
   d. Loading coils

4. As frequency increases, capacitive reactance drops. True/False

5. A lump series inductance added to a local loop is commonly referred to as a/an _____.
   a. repeater
   b. bridged tap
   c. loading coil
   d. amplifier

6. The H loading coil designation indicates a coil is added every _____ feet on the local loop.
   a. 3000
   b. 4500
   c. 6000
   d. 8000

7. The B loading coil designation indicates a coil is added every _____ feet on the local loop.
   a. 3000
   b. 4500
   c. 6000
   d. 8000

8. The D loading coil designation indicates a coil is added every _____ feet on the local loop.
   a. 3000
   b. 4500
   c. 6000
   d. 8000

9. H44 loading coils are commonly used in the United States for _____ on the local loop.
   a. data circuits
   b. local loops
   c. program networks
   d. cellular phones

10. B22 loading coils are commonly used in the United States for _____ on the local loop.
   a. data circuits
   b. local loops
   c. program networks
   d. cellular phones

11. H88 loading coils are commonly used in the United States for _____ on the local loop.
   a. data circuits
   b. local loops
   c. program networks
   d. cellular phones

12. Loaded local loops effectively block frequencies above _____ Hz.
   a. 30
   b. 300
   c. 3000
   d. 4000

13. An unterminated wire pair sitting in parallel to the local loop is commonly referred to as a/an _____.
   a. repeater
   b. bridged tap
   c. loading coil
   d. amplifier

14. Bridged taps typically cause major problems at voice transmission frequencies and must always be removed. True/False

15. A _____ converts analog signals to digital signals.
   a. multiplexer
   b. CODEC
   c. UART
   d. BIOS

16. A _____ combines multiple digital signals into a single signal that can be sent out over a single connection.
   a. multiplexer
   b. CODEC
   c. UART
   d. BIOS

17. CODECs use _____ to convert analog signals to digital signals.
   a. multiplexers
   b. PAM
   c. B8ZS
   d. PCM

18. CODECs use sampling to generate a _____ signal.
   a. multiplexer
   b. PAM
   c. B8ZS
   d. PCM

19. Nyquist's sample theorem states that, when converting an analog signal to a digital signal, the analog signal must be sampled at a rate equal to or greater than _____ its highest frequency component for analog signal reconstruction.
   a. 2 times
   b. 4 times
   c. 6 times
   d. 8 times

20. For minimal analog-to-digital and corresponding digital-to-analog conversion, an analog voice telephone line must be sampled minimally _____ times per second.
   a. 30
   b. 3300
   c. 4400
   d. 6600

21. The PCM sample rate selected by the telephone companies for analog voice telephone lines is _____ samples per second.
   a. 8000
   b. 5000
   c. 4000
   d. 300

22. In addition to sampling rate, _____ is used to generate a PCM wave.
   a. quantization
   b. induction
   c. encoding
   d. companding

23. _____ is used to compress and divide lower-amplitude voice signals into more voltage levels and provide more signal detail at lower-voltage amplitudes.
    a. quantization
    b. induction
    c. encoding
    d. companding

24. The process of giving a signal an eight-bit binary code is referred to as _____.
    a. quantization
    b. induction
    c. encoding
    d. companding

25. A _____ takes 96, 64-kilobits-per-second voice or modem signals, converts them to digital, and multiplexes them at a remote terminal.
    a. UART
    b. SLC-96
    c. T-1
    d. DS-1

Section 3.2

26. Analog or frequency multiplexing is now widely used in the United States. True/False

27. Digital multiplexing is now widely used in the United States. True/False

28. Using _____, each connected device is assigned a time slot whether or not the device has anything to send.
    a. WDM
    b. STDM
    c. TDM
    d. FDM

29. Using _____, time slots are dynamically assigned and, if a device is idle, it will not receive any time slots.
    a. WDM
    b. STDM
    c. TDM
    d. FDM

30. Any information transmitted that is not data in a communications systems is called _____.
    a. a slot
    b. a buffer
    c. a period
    d. overhead

31. _____ uses wavelength to represent different communications channels.
    a. WDM
    b. STDM
    c. TDM
    d. FDM

Section 3.3

32. A single analog voice channel, after conversion from analog to digital, requires _____ of digital bandwidth.
    a. 32 Kbps
    b. 64 Kbps
    c. 128 Kbps
    d. 256 Kbps

33. The basic building block or channel for the existing digitally multiplexed T-carrier system in the United States and for the E-carrier system in Europe is referred to as _____.
    a. DS-0
    b. DS-1
    c. DS-2
    d. DS-5

34. The DS-0 bit rate can be calculated at _____.

a. 64 Kbps
b. 6.312 Mbps
c. 560.16 Mbps
d. 1.544 Mbps

35. The T-1 bit rate can be calculated at _____.

a. 64 Kbps
b. 273.176 Mbps
c. 560.16 Mbps
d. 1.544 Mbps

36. The T-2 bit rate can be calculated at _____.

a. 43.736 Mbps
b. 6.312 Mbps
c. 560.16 Mbps
d. 1.544 Mbps

37. The T-3 bit rate can be calculated at _____.

a. 43.736 Mbps
b. 273.176 Mbps
c. 560.16 Mbps
d. 1.544 Mbps

38. The T-4 bit rate can be calculated at _____.

a. 560.16 Mbps
b. 6.312 Mbps
c. 273.176 Mbps
d. 64 Kbps

39. The T-5 bit rate can be calculated at _____.

a. 64 Kbps
b. 6.312 Mbps
c. 560.16 Mbps
d. 1.544 Mbps

Section 3.4

40. The E-1 bit rate can be calculated at _____.

a. 139.264 Mbps
b. 8.448 Mbps
c. 565.148 Mbps
d. 2.048 Mbps

41. The E-2 bit rate can be calculated at _____.

a. 33.368 Mbps
b. 8.448 Mbps
c. 565.148 Mbps
d. 2.048 Mbps

42. The E-3 bit rate can be calculated at _____.

a. 139.264 Mbps
b. 8.448 Mbps
c. 565.148 Mbps
d. 33.368 Mbps

43. The E-4 bit rate can be calculated at _____.

a. 139.264 Mbps
b. 33.368 Mbps
c. 565.148 Mbps
d. 2.048 Mbps

44. The E-5 bit rate can be calculated at _____.

a. 139.264 Mbps
b. 8.448 Mbps
c. 565.148 Mbps
d. 2.048 Mbps

Section 3.5

45. The base SONET standard OC-1 bit rate is _____.

a. 1.544 Mbps
b. 51.84 Mbps

c. 2.48832 Gbps

d. 9.95328 Gbps

46. The SONET standard OC-48 bit rate is
_____.

   a. 1.544 Mbps

   b. 51.84 Mbps

   c. 2.48832 Gbps

   d. 9.95328 Gbps

47. The SONET standard OC-192 bit rate is
_____.

   a. 1.544 Mbps

   b. 51.84 Mbps

   c. 2.48832 Gbps

   d. 9.95328 Gbps

Section 3.6

48. T-1 carriers use a bit in the sixth and twelfth
frames of every T-1 carrier for in-band signaling. This process is called _____.

   a. bit borrowing

   b. bit biting

   c. bit taking

   d. bit robbing

49. T-1 carriers are limited to _____ bits per
frame in all frames.

   a. 4

   b. 5

   c. 6

   d. 7

50. _____ is used to add synchronization on
"quiet" T-1 lines.

   a. CODEC

   b. B8ZS

   c. SONET

   d. Ethernet

Section 3.7

51. Packet-switched networks are designed to
use dedicated circuit line connections like
the PSTN. True/False

52. Data traffic is commonly referred to as
_____.

   a. sluggish

   b. fast

   c. bursty

   d. flashy

53. Frame relay does not use standard bit rates.
True/False

54. The frame-relay _____ takes data from a
customer and breaks it down into frames.

   a. PAD

   b. PSE

   c. ATM

   d. frame

55. The frame-relay _____ routes frames
through the frame-relay cloud.

   a. PAD

   b. PSE

   c. ATM

   d. frame

56. Frame relay uses _____ as network connections within the frame-relay cloud.

   a. PADs

   b. PVCs

   c. ATMs

   d. multiplexers

57. Extra frame-relay address and control information added to the frame is called
_____.

   a. PAD

   b. PSE

   c. PVC

   d. overhead

Section 3.8

58. ATM is considered connection-oriented. True/False

59. ATM uses _____, fixed-length cells.
    a. 8-bit
    b. 53-bit
    c. 53-byte
    d. 155.52-megabyte

60. The ATM cell header is _____ bytes long.
    a. 2
    b. 3
    c. 5
    d. 48

61. The ATM cell payload is _____ bytes long.
    a. 2
    b. 3
    c. 5
    d. 48

62. All ATM error checking and correction is done within the ATM cloud. True/False

Discussion Questions

Section 3.1

1. A local loop designed for voice is 6000 feet long. Calculate the capacitive reactance for the loop at:
   a. 1 KHz.
   b. 2 KHz.
   c. 3 KHz.

2. A local loop for voice is 18,000 feet long. Calculate the capacitive reactance for the loop at:
   a. 1 KHz.
   b. 2 KHz.
   c. 3 KHz.

3. Refer to your answers in problems 1 and 2. What happens to the capacitive reactance of the loop when frequency remains constant but loop length changes? Be sure to refer to each of the three individual frequencies used in your calculations.

4. Refer to your answers in problems 1 and 2. What happens to the capacitive reactance of the loop when loop length remains constant but frequency changes? Be sure to refer to each of the three individual frequencies used in your calculations.

5. Calculate the cutoff frequency and sketch a frequency response curve for a local 12,000-foot loop using H88 coils spaced every 6000 feet.

6. Calculate the cutoff frequency and sketch a frequency response curve for a local 18,000 foot loop using H88 coils spaced every 6000 feet.

7. Compare your answers in problems 5 and 6. Are they the same? Why or why not?

8. Explain why a bridged tap typically will not cause problems at lower voice frequencies.

9. Describe how a CODEC works to convert analog signals to digital and also to convert digital signals to analog.

10. A system similar to the PSTN has a bandwidth of 9500 Hz. Using the Nyquist sam-

pling theorem, calculate the minimal sampling rate.

11. Explain the process of quantization.

12. Describe how and why companding is used.

Section 3.2

13. Why is frequency multiplexing no longer used in the United States?

14. One frequency multiplexed jumbogroup multiplex is the equivalent of how many jumbogroups? mastergroups? supergroups? groups?

15. Contrast TDM and STDM. Is one more efficient than the other? Why or why not?

Section 3.3

16. Explain why a single analog voice channel, after conversion from analog to digital, requires 64 Kbps of digital bandwidth.

17. One multiplexed T-5 carrier is the equivalent of how many T-4 carriers? T-3 carriers? T-2 carriers? T-1 carriers? DS-0 channels?

Section 3.4

18. One multiplexed E-5 carrier is the equivalent of how many E-4 carriers? E-3 carriers? E-2 carriers? E-1 carriers? DS-0 channels?

Section 3.5

19. One SONET OC-192 carrier is the equivalent of how many OC-48 carriers? OC-1 carriers? DS-0 channels?

Section 3.6

20. Describe the term *56-Kbps clear channel capacity* with reference to T-1 carrier, single-channel, maximum-data bit rate. Reference your answer to the fact that one DS-0 channel has a value of 64 Kbps.

21. Explain why T-1 carrier signals alternate from positive to negative.

22. Explain the process of B8ZS and how it adds synchronization to quiet T-1 lines.

Section 3.7

23. Explain why frame-relay networks are commonly represented using a cloud.

24. What is meant by the term *bursty?* Explain how this term applies to data traffic.

25. Explain why frame-relay PVCs are called *virtual.*

26. Describe the frame-relay frame structure. If the data section is at a maximum value of 1610 bytes, is transmission more efficient than a frame with a smaller data section? Explain.

Section 3.8

27. What are the two common ATM bit rates? Use the Internet to determine if other bit rates are possible. If so, explain.

28. Explain the advantages of using a fixed cell size for ATM.

29. Contrast ATM's fixed cell size with frame relays' variable frame size. Is one faster than the other? Explain.

Transmission Media

Objectives Upon completion of this chapter, the student should be able to:

- Describe various types of telecommunications transmission media.
- Discuss the disadvantages of open copper wire systems.
- Discuss the advantages of insulated copper wire systems.
- Model a transmission system in block diagram format.
- Describe wire color codes.
- Define crosstalk.
- Describe various types of copper wire and list advantages and disadvantages of each.
- Describe coaxial cable and list advantages and disadvantages.
- Describe various types of wireless systems and list advantages and disadvantages of each.
- Describe various types of fiber-optic cable and list advantages and disadvantages of each.

Outline 4.1   Copper Systems
4.2   Wireless Long-Haul Systems
4.3   Optical Fiber Systems

Key Terms

binder
coaxial cable
crosstalk
dispersion

Electronics Industry Association/ Telecommunications Industry Association (EIA/TIA)

Ethernet
far-end crosstalk (FEXT)
geosynchronous earth orbit

graded-index fiber	microwave	shielded-twisted-pair (STP) cable
index of refraction	near-end crosstalk (NEXT)	single-mode fiber
inside wire (IW)	network interface (NI)	step-index-fiber
insulators	outside wire	super-unit binder
laser	point of demarcation	unshielded-twisted-pair (UTP) wire
laser diode	satellite	
L-carrier		
local-area network (LAN)		

Introduction

Transmission media can be loosely categorized into three separate areas: copper, optical fiber, and wireless. This chapter addresses each of the three.

4.1 Copper Systems

Copper media was first used by Bell for voice transmission and is still widely used today. The initial voice network was built in the United States using open, or uninsulated, wire. As time went on, the need for higher bandwidth increased and copper twisted pair and coaxial cable were used.

Open Wire

Open wire, which was popular until the 1970s in rural communities across the United States, is uninsulated thick copper wire strung from telephone pole to telephone pole. Open-wire systems were commonly used for the local loop and connected residences and businesses to the local telephone company switching-station central office, as illustrated in figure 4-1.

The wires were strung telephone pole to telephone pole and were attached to the poles with glass **insulators.** The insulators were used to isolate the open wires from ground on the telephone poles, as illustrated in figure 4-2.

The wires, insulators, and poles were all exposed to weather extremes, including heat and cold. The glass insulators had a tendency to get wet and collect dust and dirt. Over a period of time, the grime would thicken, coating the insulators and making contact with the grounded telephone pole. During wet weather, the grime would get wet and make conductive contact with the grounded telephone pole, as shown in figure 4-3. Transmission signals (voice) would lose strength and sometimes even disappear, leaking through this conductive path. Lightning strikes also caused serious problems to

FIGURE 4-1 Open-wire local loop.

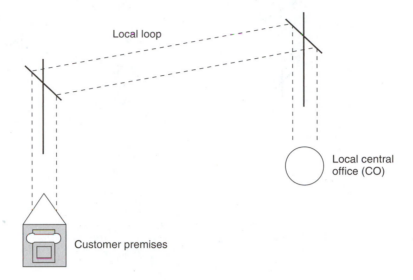

FIGURE 4-2 Early AT&T open-wire glass insulator.

highly susceptible conductive open-wire systems. Special ceramic lightning arresters were required wherever equipment was attached.

Heavy-gauge, open-wire systems were expensive to install and maintain. The maximum number of voice channels that could be transmitted on a single, open-wire pair was 12; and, as neighborhoods grew, the wire space on the telephone poles ran out. Most open-wire systems in the United States have now been replaced.

FIGURE 4-3 Open-wire system.

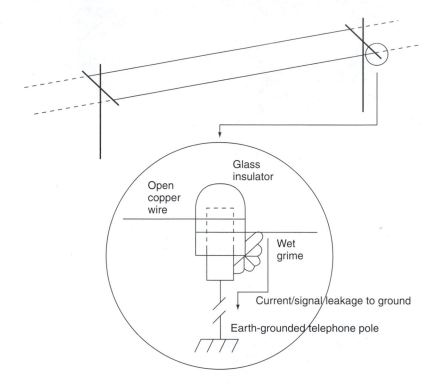

Insulated Copper Wire

The solution to the open-wire-pair limitations was to use insulated wire and bundle wires together. Today, a typical local loop in the United States is 3 miles (approximately 18,000 feet) of 22 AWG twisted-pair copper wire. This 22 AWG wire carries a resistance specification of 19 ohms per 1000 feet at 77° F (25° C) along with specifications for line inductance, capacitance, and conductance.

In chapter 3, we learned that wire resistance is directly proportional to wire gauge. In other words, as wire gauge increases, wire resistance also increases. On the local loop, we are very concerned with DC wire resistance. The object of any transmission system is to deliver as much input (CO) power as possible to the output device (telephone at the end of the loop). This is illustrated in figure 4-4 in block-diagram format.

Multistrand, polyurethane-plastic-insulated wire provides good insulation from ground contact and, with the right cable specification, can be buried, further protecting the lines from lightning strikes and other damaging effects.

Point of Demarcation

Most people are familiar with the "box" that a telephone company technician has installed on or in their homes. This box is referred to as a **network**

FIGURE 4-4
Transmission system
block diagram.

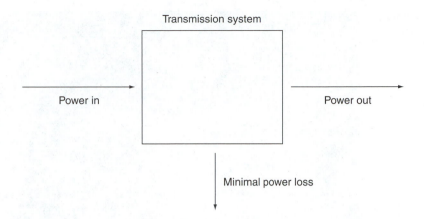

FIGURE 4-5 Point
of demarcation.

interface (NI) and is installed by a telephone company technician. The network interface indicates the **point of demarcation** on the telephone network that defines or differentiates between installing telephone company equipment and the customer equipment. Figure 4-5 indicates the separation between these two networks.

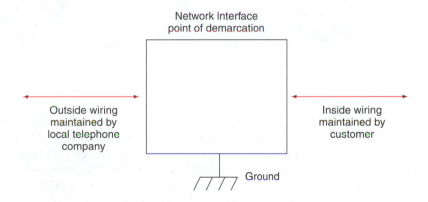

The network interface is used to separate what is called **outside wire**, which the telephone company installs, connects to, and maintains; and **inside wire (IW),** commonly referred to as POTS or JK wire which the customer installs and maintains. The network interface must be grounded for protection against electrical spikes and surges caused by things like lightning. A typical residential network interface is shown in figure 4-6. Both inside and outside wires are bundled together and are identified using different colors that correspond to an industry-specified color code.

Inside-Wire (IW) Industry Color Code

Most network interfaces have a terminal strip for connecting inside telephone wires and an RJ-11 telephone jack for plugging in a test telephone.

FIGURE 4-6 Typical residential network interface.

Most telephones only require two wires to be connected to the network interface for operation. These wires are referred to as the *tip* and the *ring* by most technicians. Plain old telephone system (POTS) wire—also referred to as direct inside wire (DIW), station, or "JK" wire—is typically run through most homes in the United States. "Normal" inside wire has four wires that combine to form two wire pairs, as shown in figure 4-7. The four bundled wires are colored (red, green, black, yellow), with the wire color combinations indicating the two pairs.

Typically, the first pair uses the green wire (*tip*) and the red wire (*ring*) and the second pair uses the black wire (*tip*) and the yellow wire (*ring*). When a single phone line is used, only the red and green wires are used and the black and yellow are extras that can be used later for a second phone line. The inside-wire color code is illustrated in figure 4-8.

Cable Color Code

The telephone industry has developed a color code for wire bundles with more than two pairs. This code consists of the primary colors: blue, orange, green, brown, and slate (gray); and the secondary colors: white, red, black,

FIGURE 4-7 Typical residential POTS wire.

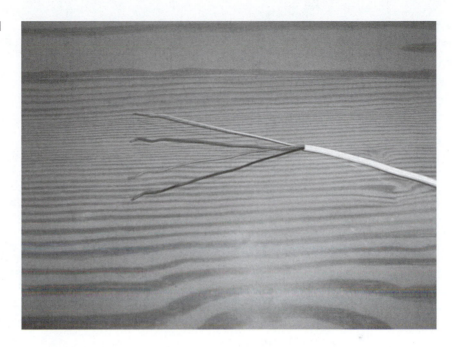

FIGURE 4-8 Inside-wire color code.

Tip — Green
Ring — Red
} Phone line 1

Tip — Black
Ring — Yellow
} Spare or phone line 2

FIGURE 4-9 Wire-bundle wire colors.

Primary colors	Secondary colors
Blue	White
Orange	Red
Green	Black
Brown	Yellow
Slate	Violet

yellow, and violet. These colors are indicated in figure 4-9. Here, the tip wire is commonly the secondary color with marks of the primary color—for example, orange with white marks or blue with white marks. The ring wire is commonly the primary color, with marks of the secondary color—for example, white with orange marks or white with blue marks.

FIGURE 4-10 Wire color-code hierarchy.

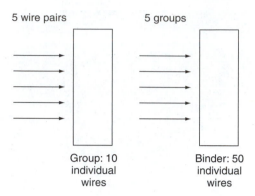

These wire combinations are put into *groups*. A group of wires consists of five wire pairs, and a **binder** consists of five combined wire groups. Group and binder combinations are illustrated in figure 4-10.

We can continue with combining wire pairs. In a 25-pair binder, 5 groups of 5 wire pairs are formed as follows:

Group 1 Pairs

Blue white/white blue

orange white/white orange

green white/white green

brown white/white brown

slate white/white slate

Group 2 Pairs

blue red/red blue

orange red/red orange

green red/red green

brown red/red brown

slate red/red slate

Group 3 Pairs

blue black/black blue

orange black/black orange

green black/black green

brown black/black brown

slate black/black slate

FIGURE 4-11
Binder/super-unit
binder hierarchy.

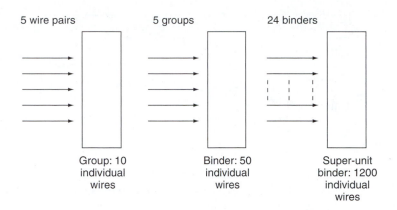

Group 4 Pairs
blue yellow/yellow blue
orange yellow/yellow orange
green yellow/yellow green
brown yellow/yellow brown
slate yellow/yellow slate

Group 5 Pairs
blue violet/violet blue
orange violet/violet orange
green violet/violet green
brown violet/violet brown
slate violet/violet slate

Binders

Cables with more than 25 pairs use different colored threads—also called binders—to identify the combined groups. Each binder contains five groups with the same group color combinations repeated in each binder. For cables with more than 24 binders (600 pairs), 24 binders at a time are combined into **super-unit binders,** as illustrated in figure 4-11.

Twisted-Pair, Data-Grade Cabling

Most new homes and businesses are no longer having internal station wire installed. They are having data-grade, four-pair **unshielded-twisted pair (UTP) wire** installed. Unshielded-twisted pair (UTP) wire is copper wire

TABLE 4-1
UTP Standards.

Cable category	Number of pairs	Maximum-signal bandwidth
Category 3	4	16 MHz
Category 4	4	20 MHz
Category 5	4	>100 MHz
Category 6	4	>200 MHz

commonly used for **local-area-network (LAN)** construction. With the evolution of local-area networks, copper wire technology has improved and now supports some very high data rates. This same data-grade UTP wiring can transmit voice as well as data. The **Electronics Industry Association/ Telecommunications Industry Association (EIA/TIA)** is a dual organization that is accredited by the American National Standards Institute (ANSI) and has been set up to develop voluntary industry standards for a wide variety of telecommunications products. The EIA/TIA has developed the 568 unshielded-twisted-pair standards that now include three different UTP data cable categories. These categories are shown in table 4-1. Each of these UTP cable categories can be used for both voice and data transmission with Category 5 the most commonly used in new construction for both data and voice. On June 20, 2002 the higher bandwidth Category 6 standard, also known as the TIA/EIA-568-B.2-1 standard was approved. The Category 6 standard is backward compatible with Categories 3, 4 and 5. A category 5 UTP cable with RJ-45 Connector end connector is shown in figure 4-12.

FIGURE 4-12
Category 5 UTP wire with RJ-45 connector.

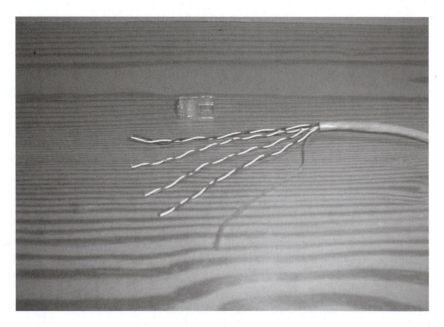

TABLE 4-2
Category 5 cable
translations.

Category 5 color	POTS/station wire equivalent
Blue	Red
White-blue	Green
White-orange	Black
Orange	Yellow

The color code on category 5 cable is different than the station wire described earlier in this chapter. Category 5 cable uses the following wire combination colors: blue, white-blue, orange, white-orange, green, white-green, brown, white-brown. When mixed with station wire or when installing in place of station wire, the translations shown in table 4-2 are commonly used.

Crosstalk

The twisting of the pairs is critical in the elimination of **crosstalk** between adjacent wires. Crosstalk occurs when electrical noise caused by electromagnetic fields on one conductor is electromagnetically coupled onto another conductor in close proximity. Electrical current flowing through any conductor will produce a surrounding electromagnetic field. If another conductor is within the surrounding field, an inductively coupled current will flow through the adjacent conductor. This is illustrated in figure 4-13. In the figure, current flowing through the conducting wire will produce an inductively coupled current in the adjacent wire. If the varying-signal current represents a voice transmission, the conversation can cross over from one line to another and voices can be heard on one line from another line conversation. Usually, this is only an annoyance since crosstalk signal levels are typically low when compared with the signal levels of the conversation on the primary line. On the other hand, digital data transmissions are extremely sensitive to crosstalk. Crosstalk can cause bit misinterpretation and will typically require a retransmission of the damaged data.

FIGURE 4-13
Inductively coupled
electromagnetic flux.

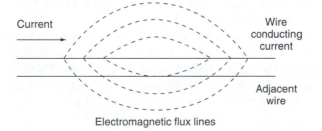

FIGURE 4-14 Near-end crosstalk.

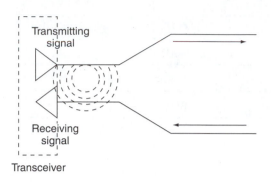

FIGURE 4-15 Far-end crosstalk.

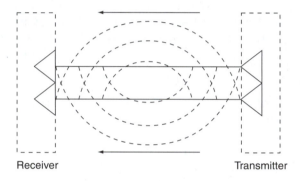

Near-End Crosstalk (NEXT)

Near-end crosstalk (NEXT) is crosstalk that occurs between a transmitted signal and a received signal. Transmitted signals are typically stronger than signals that are being received and interfere with the received signals. Near-end crosstalk (NEXT) is illustrated in figure 4-14.

Far-End Crosstalk (FEXT)

Far-end crosstalk (FEXT) is crosstalk that occurs between two signals transmitted in the same direction. The adjacent conductors each produce a magnetic field and can interfere with each other. Far-end crosstalk (FEXT) is illustrated in figure 4-15.

The most common way to reduce crosstalk between adjacent wires is to twist the wires together in a way that cancels the crosstalk flux. **Shielded-twisted pair (STP) cable** includes additional shielding in the form of foil or metallic braid; this additional shielding protects against crosstalk and other forms of electromagnetic interference.

FIGURE 4-16
Coaxial cable cross section.

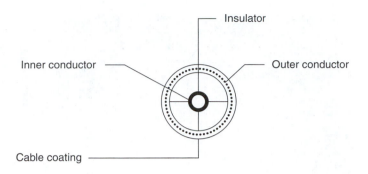

Coaxial Cable

As the use of the telephone grew, the channel capacity of copper wire did not have the bandwidth required to interconnect telephone centers over long distances. **Coaxial cable** was commonly used for these early copper wire long-distance connections but has now been replaced with higher bandwidth capacity transmission media like microwave and fiber. A coaxial cable cross-sectional illustration is shown in figure 4-16. Coaxial cable has an inner core used for signal transmission and an outer core that is either a metallic braid or foil. The outer core acts as an interference signal shield.

The insulator between the inner and outer cores is typically made of either mica or plastic, and the cable coating is either polyurethane or Teflon™. Teflon cable coating is used on plenum-grade cable that is used when fire regulations require it. Plenum-grade cable does not produce toxic fumes when it burns, while polyurethane cable does. Common coaxial cable terminators and connectors are shown in Figure 4-17.

Coaxial cable first replaced open-wire connections between telephone centers. It offers higher bandwidth and higher protection from the elements, and it can be buried. Transmission bandwidth depends on the type of insulator used in the cable, with higher-quality insulator cable offering higher potential bandwidths.

L-Carrier

The **L-carrier** system is an obsolete system developed by AT&T in the 1970s. This system used coaxial cable to carry multiplexed voice conversations from telephone center to center.

The L-carrier system had several versions, with the most popular being the L5E first introduced in the 1970s. The L5E consists of a bundle of 11 coaxial cable pairs. One pair was considered a spare, and 10 were used for transmission. Each individual pair could carry 13,200 simultaneous voice conversations allowing 132,000 simultaneous voice conversations to be transmitted from telephone company center to center over one L5E cable.

FIGURE 4-17
Coaxial cable with
connectors.

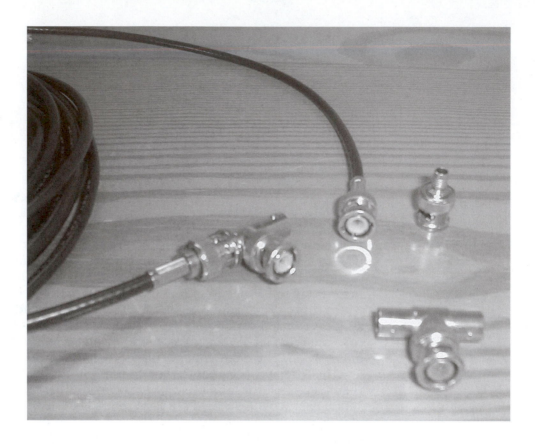

In the early 1980s computer networking became popular and coaxial cable was used to construct the first ARCNet and **Ethernet** local-area networks (LANs). In addition, many cable telephone companies use coaxial cable to deliver television to their customers.

4.2 Wireless Long-Haul Systems

As the telephone network grew, long-haul wireless systems became popular. Depending on the frequency chosen, wireless systems can have higher noise immunity than cable systems and are typically more simple to set up and maintain when compared to land-based cable systems. Wireless systems do not require cable to be strung or buried and do not require a right of way. Wireless systems can also be more quickly modified and moved when compared to cable systems. To fundamentally understand wireless transmission, a brief outline of the radio spectrum is required.

The Radio Spectrum

The *radio spectrum* is defined as the part of the spectrum that can be used for communication purposes. In the United States, this range of frequencies is divided into pieces called *frequency bands* by the Federal Communications Commission (http://www.fcc.gov). Information on how the radio spectrum is used in the United States can be found on the FCC Web site at: http://www.fcc.gov/oet/info/database/spectrum/Welcome.html

The FCC controls U.S. frequency allocation but does not control allocation in other countries. Some allocations are considered worldwide, while others are limited to specific regions and countries. The International Telecommunications Union (ITU, http://www.itu.int), headquartered in Geneva, Switzerland, is an international organization within which governments and the private sector coordinate global telecommunications networks and services. Included in these standards is an international table of frequency allocations maintained by the *World Radiocommunication Conference* members. The regulated frequency band ranges from 9 kilohertz to 400 gigahertz and is segmented into smaller bands and allocated to over 40 defined radiocommunication services.

Radio regulations contain the international table of frequency allocations (currently included in ITU Article S5), which is based on a block allocation method, with footnotes. The regulated frequency band (9 kilohertz to 400 gigahertz) is segmented into smaller bands and allocated to over 40 defined radiocommunication services.

All member countries and organizations that are part of the ITU are invited to participate as part of the conference group. Each frequency block is defined as either primary or secondary, with a secondary designation not interfering with a primary designation. Amendments to the table are made after discussion and agreement between the international groups. If a country wants to use a frequency band or block that is different from what the rest of the conference group has agreed to, a vote is taken among members and, if there is consensus, a footnote is added to the international table of frequency allocations.

Microwave

As bandwidth requirements have grown, most of the coaxial cable connecting telephone company centers were replaced by higher-bandwidth transmission systems using **microwave** methods. Microwave is defined as wireless electromagnetic energy having a frequency higher than 1 gigahertz.

FIGURE 4-18
Microwave
transponder system
block diagram.

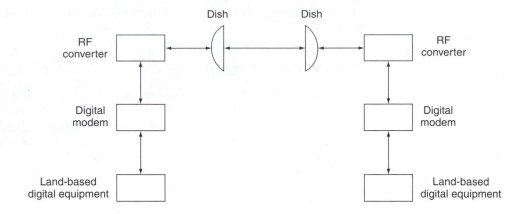

Microwave radio transponder systems are used to transmit digitized voice, data, and video between two line-of-sight locations at distances up to approximately 60 kilometers. The microwave radio transponder consists of a *digital modem* that interfaces with land-based digital equipment; a *radio frequency (RF) converter,* which is a device used to convert digital modem signals to microwave frequencies; and a *dish antenna* to transmit and receive signals. Two microwave transponder systems communicating with each other are referred to as a *microwave hop,* as indicated in figure 4-18.

Two microwave frequency bands were first used: the 4-gigahertz band (3.7–4.2 gigahertz) and the 6-gigahertz band (4.924–6.425 gigahertz). These bands were first chosen because water caused by rain does not have as large a fading effect as it does at higher microwave frequencies. Each band was split into channels with the 4-gigahertz band having channel bandwidth of 20 megahertz and the 6-gigahertz band having channel bandwidth of 30 megahertz. Today, higher-capacity bands including 11 gigahertz and 18 gigahertz are used. Microwave systems are expensive to install and maintain but work great for wireless line-of-sight communications. Transmitters and receivers can be placed up to 60 kilometers apart as long as they are within sight of each other. Microwave systems are also used for digital system interconnection between buildings producing data rates of over 2 megabits per second.

Satellite Systems

Satellite systems are used to carry telephone and television traffic to different parts of the world. Satellites are combination transmitters and receivers that are launched into space and set in orbit around the earth. Satellite systems use earth station ground dishes to communicate with satellites in **geosynchronous earth orbit.** A geosynchronous earth orbit satellite is a satellite that rotates at the same rate as the earth revolves around its axis. In geosynchronous orbit, a satellite appears *stationary* to earth stations that use it. This is illustrated in figure 4-19.

FIGURE 4-19
Geosynchronous
Earth orbit satellite.

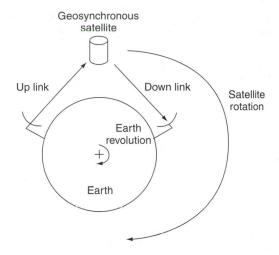

TABLE 4-3 Satellite
communication
frequency bands.

Band	Frequency range
C band	3 GHz–6 GHz
X band	7 GHz–9 GHz
Ku band	18 GHz–22 GHz

Satellites have *transponders* that are powered by solar cells, and each transponder is responsible for a different satellite frequency band. The transponders take uplinked signals that, by the time they get to the satellite, are very weak and use low-noise amplifiers to amplify these signals. Once the uplink signals are amplified, they are translated to a different frequency and retransmitted by the satellite to the downlink area.

The ITU has reserved space in the 2.4 to 22-gigahertz frequency spectrum for satellite transmissions. These are microwave frequencies, and the same rules apply to satellite uplinks and downlinks. An earth station must have a direct line of sight for satellite communication to work.

Several popular frequency bands are used for satellite communication, including those listed in table 4-3. In the 1960s, the first communication satellite systems used the C-band frequency range of 3.7 to 4.2 gigahertz for the downlink transmission and the frequency range of 4.925 to 6.425 gigahertz for the uplink transmission. At that time, land-based terrestrial microwave systems were also using these same frequencies, which could have caused a problem if the land-to-land-based dishes were to receive a significantly strong satellite signal. For this reason, satellite power levels were restricted to a level that would not cause terrestrial microwave link interference.

In the late 1970s, communications satellites shifted to the Ku frequency band where little terrestrial microwave communications were done. This allowed the satellites to transmit more powerful signals than the C-band systems. We can use the cable television system as an example. For good C-band satellite television reception, a dish of 2 to 3.7 meters is required. Ku-band satellites transmit at higher power levels and can use receiving dishes as small as 30 centimeters in diameter for the same quality reception. There is a trade-off, however. The higher frequency and corresponding shorter wavelength Ku band produces a beam of narrower width than the lower frequencies. This reduces earth beam footprint coverage. In addition, at frequencies above 10 gigahertz, clouds, rain, and snow can significantly reduce downlink signal levels.

4.3 Optical Fiber Systems

In 1966, scientists at ITT first used *fiber optics*—commonly referred to as *fiber* in the telecommunications industry—as a transmission medium for electrical signals in the lab. By 1980, AT&T had installed the first long-haul fiber link from Cambridge, Massachusetts, to Washington, D.C. Today, fiber is widely used in most parts of the world, providing high-bandwidth transmission over long distances. A piece of communications fiber is shown in figure 4-20.

FIGURE 4-20
Optical cable showing fiber, KEVLAR[tm] strain relief, and insulation.

Optical fiber systems have several advantages over copper or wireless systems. They provide higher bandwidths, are relatively immune to electromagnetic interference, are lightweight, are small in size, and are much more secure. An optical telecommunications system, as illustrated in figure 4-21, consists of a *transmitter* that converts electrical signals to light, *optical fiber* that transmits the light signals, and a *receiver* that converts the light signals back to electrical signals.

Optical fiber can be made from either glass or plastic, with the higher-bandwidth products made with glass. Sources are typically **lasers** in the form of **laser diodes.** Lasers are devices that produce light that is one specific frequency and is polarized. *Laser diodes* are miniature versions of lasers that also produce polarized light that is one specific frequency. The three major fiber types used in the telecommunications industry are step-index, graded-index, and single-mode fibers.

Step-Index Fiber

Step-index fiber was the first communications fiber developed and used. It has relatively low bandwidth capacity and high loss when compared with other more modern fiber types. A step-index fiber consists of a glass core surrounded by a glass cladding that has a lower **index of refraction** than the core.

The *index of refraction* of a material is the ratio $c:v$, where c is the speed of light in a vacuum and v is the speed of light in the specified material. Light speed in a vacuum is at a maximum rate; therefore, light will always travel more slowly in any other transmission medium. This means that any transmission medium, besides a vacuum, will have an index of refraction of greater than 1.

As light moves down a piece of step-index fiber, it moves slower through the core than through the cladding due to the difference in the index of refraction. This process causes the light in the core to be internally reflected and to pass down the core, as illustrated in figure 4-22. Looking at the figure, we can see that light launched into a step-index fiber at more normal (perpendicular on center axis of fiber) angles travels closer to the center of the core and is referred to as *lower-order mode*. Lower-order modes end up traveling a shorter distance than the higher-order modes that travel farther away from the core center. This is illustrated in more detail in figure 4-23.

FIGURE 4-21
Optical telecom-
munications system.

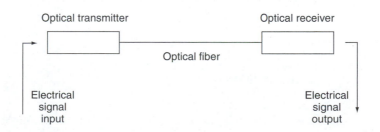

FIGURE 4-22 Step-index fiber, side view and cross-sectional view.

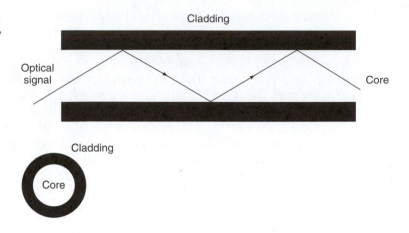

FIGURE 4-23 Lower- and higher-order modes.

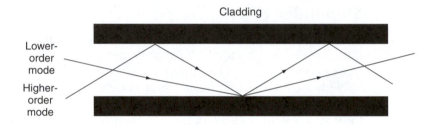

This separation of light waves as they travel down a piece of fiber-optic cable is defined as modal **dispersion,** and step-index fiber suffers greatly from it. Values of 20 to 30 nanoseconds per kilometer are common. For this reason, step-index fiber typically is not used as a telecommunications transmission medium.

Graded-Index Fiber

Graded-index fiber is optical fiber that corrects the modal dispersion problems that step-index fiber has by using a graded, refractive-index core. Graded-index fiber is manufactured so that the outer cladding fiber layers have a lower refractive index of refraction than the center of the core, as illustrated in figure 4-24. This layering of cladding causes the higher-order

FIGURE 4-24 Graded-index-fiber core.

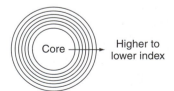

FIGURE 4-25
Graded-index-fiber
transmission.

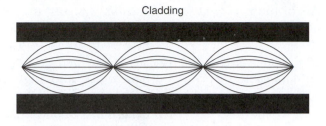

Cladding

FIGURE 4-26
Single-mode-fiber
transmission.

Cladding

Core

modes to move faster in the outer layers, keeping up with the lower-order modes traveling closer to the core center, as illustrated in figure 4-25. Modal dispersion values for graded-index fiber are typically less than 1 nanosecond per kilometer.

Single-Mode Fiber

Single-mode fiber is optical fiber that has been designed to only allow one mode to travel down the fiber core. The fiber is constructed with a core diameter of the specific wavelength of the optical signal being used, and the cladding is constructed to be thicker than 10 times the core diameter to minimize modal dispersion. Typical wavelengths used are 1300 and 1500 nanometers, and these match corresponding core diameter values. Single-mode-fiber transmission is illustrated in figure 4-26.

Single-mode fibers have potential bandwidths of greater than 100 gigahertz. Today, long-haul optical communications links are composed of primarily single-mode fiber.

Summary

1. *Copper media* was first used by Bell for voice transmission and is still commonly used today.

2. *Open-wire systems* were common in the rural areas of the United States until the 1970s.

3. The *typical local loop today* is 3 miles of 22 AWG twisted-pair copper wire.

4. The *network interface* box installed by a telephone company technician is considered the *point of demarcation* between the installing telephone company and the customer.

5. The *inside wire* industry color code consists of green, red, black, and yellow wires.

6. The *cable color code* consists of multicolored wires bundled using a standard hierarchical color-code system.

7. The *EIA/TIA* has developed the 568 standard for unshielded-twisted-pair data cable.

8. *Crosstalk* occurs when current is inductively coupled from one conductor to another.

9. *Coaxial cable* was once commonly used for high-bandwidth connections but has now been replaced with fiber and wireless systems.

10. *Microwave systems* can be used to transmit digitized voice, data, and video between two line-of-sight locations at distances greater than 60 kilometers.

11. *Satellite systems* use earth station ground dishes to communicate with satellites in geosynchronous orbit.

12. *Step-index optical fiber* has relatively low bandwidth and high loss when compared with other fiber types.

13. *Graded-index optical fiber* has low bandwidth and higher loss than single-mode fiber.

14. *Single-mode optical fiber* has potential bandwidths of greater than 100 gigahertz and is used almost exclusively for long-haul communications.

Review Questions

Section 4.1

1. Open-wire systems are no longer used by telephone companies in the United States. True/False

2. The maximum number of voice channels that could be transmitted on a single, open-wire pair was _____.

   a. 1
   b. 4
   c. 10
   d. 12

3. The point of demarcation between the installing telephone company and the customer is typically the _____.

   a. telephone pole
   b. pedestal box
   c. network Interface
   d. telephone handset

4. Inside wire typically comes with two pairs (four wires). Which wire is commonly used as the first pair *tip?*

   a. Black
   b. Green

   c. Red
   d. Yellow

5. Inside wire typically comes with two pairs (four wires). Which wire is commonly used as the first pair *ring?*

   a. Black
   b. Green
   c. Red
   d. Yellow

6. Inside wire typically comes with two pairs (four wires). Which wire is commonly used as the second pair *tip?*

   a. Black
   b. Green
   c. Red
   d. Yellow

7. Inside wire typically comes with two pairs (four wires). Which wire is commonly used as the second pair *ring?*

   a. Black
   b. Green

c. Red

d. Yellow

8. A cable color-code group consists of _____ wire pairs.

a. 1

b. 3

c. 5

d. 7

9. A cable color-code binder consists of _____ wire groups.

a. 3

b. 5

c. 7

d. 10

10. A cable color-code binder consists of _____ wire pairs.

a. 15

b. 20

c. 25

d. 30

11. Which unshielded-twisted-pair category cable is rated the highest?

a. 2

b. 3

c. 4

d. 5

12. Crosstalk is primarily due to _____.

a. poor plastic wire insulation

b. inductively coupled current

c. high humidity

d. wire exposure to sunlight

13. Crosstalk is most commonly reduced by _____.

a. adding more plastic wire insulation

b. keeping cables in dark conduit

c. keeping wires in cool and dry air-conditioned environments

d. twisting wires together

Section 4.2

14. In the United States the _____ controls wireless frequency band allocation.

a. FCC

b. COMPTIA

c. ITU

d. IEEE

15. Two microwave systems communicating with each other are commonly referred to as a microwave _____.

a. route

b. switch

c. hub

d. hop

16. In geosynchronous earth orbit, a satellite appears _____ to the earth stations that use it.

a. to move vertically

b. stationary

c. only during daylight hours

d. to move horizontally

17. The satellite _____ uses the 7 to 9 gigahertz frequency range.

a. C band

b. X band

c. Ku band

d. terrestrial microwave

18. The satellite _____ uses the 3- to 6-gigahertz frequency range.

a. C band

b. X band

c. Ku band

d. terrestrial microwave

19. The satellite _____ uses the 18- to 22-gigahertz frequency range.

a. C band

b. X band

c. Ku band

d. terrestrial microwave

Section 4.3

20. Plastic optical fiber typically has a higher bandwidth capacity than optical fiber made from glass. True/False

21. _____ fiber has relatively low bandwidth and high loss when compared with other fiber types.
    a. Graded-index
    b. Step-index
    c. Single-mode
    d. All of the above

22. _____ fiber has multiple layers with the outer glass layers having a lower refractive index than the layers closer to the core.
    a. Graded-index
    b. Step-index
    c. Single-mode
    d. All of the above

23. _____ fiber typically uses wavelengths between 1300 and 1500 nanometers that match core diameter values.
    a. Graded-index
    b. Step-index
    c. Single-mode
    d. All of the above

Discussion Questions

Section 4.1

1. List three disadvantages of open-wire systems.
2. Define point of demarcation.
3. What is a network interface with reference to a residential telephone customer?
4. List the EIA/TIA UTP 568 category 3, 4, and 5 cable maximum-signal bandwidth specifications.
5. A residence has a mix of station wire and category 5 UTP. Draw a diagram indicating how the station wire and UTP should be joined.
6. Define crosstalk.
7. What is the difference between near-end crosstalk and far-end crosstalk?

8. How can crosstalk be reduced?
9. List three advantages of coaxial cable.

Section 4.2

10. List three advantages of microwave transmission systems.

Section 4.3

11. List three advantages of optical fiber systems.
12. Explain how step-index fiber works.
13. Explain how graded-index fiber works.
14. Explain how single-mode fiber works.
15. Contrast step-index, graded-index, and single-mode fiber.

Switching

Key Terms

blocking	grade of service	party line
busy hour	interexchange carrier (IXC)	point of presence (POP)
century call second (CCS)	interLATA call	private-branch exchange (PBX)
crossbar switch	intraLATA call	reed switch
electronic switching system (ESS)	local access and transport area (LATA)	remendur
Erlang	load factor	step-by-step switch
exchange	operator	Strowger switch

switchboard   toll-connecting trunk   traffic density
tandem office   toll office   trunks
toll center   traffic

Introduction

Almost immediately after the invention of the telephone, Bell realized that he needed to develop a way for calls to be switched or connected from one telephone to another. For two of his 1876 telephones to be connected, a dedicated wire pair connected from telephone to telephone was required. His initial invention only allowed conversation to occur between already connected users. We now understand that the term *connected users* means a dedicated copper wire pair connecting the two user telephones. This was great if you only needed to talk with one other person who was always at the same telephone. What if you wanted to talk with someone else? An unswitched network would require that you, and anyone else you wanted to talk to, have a pair of wires coming into each premise. Each wire pair would need to be clearly labeled. You and this person you wanted to talk with would have to set up a call schedule. For example, if you wanted to talk to a specific person and you already had a wire pair coming into your home from this person, you and this person would have to agree on a call time. A few minutes before the call time, you would each have to go and find the corresponding wire pair in your premises and connect your phone to it. At the agreed-upon time, you would each pick up the phone and hope the other person remembered to do the same. Bell knew that there was no way his invention would ever become popular and widespread using this method. Think about it; if you had 100 people you wanted to talk to from time to time, you would have 100 separate pairs of wires coming into your home. Labeling, disconnecting, and connecting would be very complex and confusing tasks. You would also not be able to make random calls; every call would have to be agreed upon in advance for setup. Bell realized almost immediately that if his invention was ever to become popular and widespread in use an easier method of switching connected wire pairs had to be developed.

5.1 The First Switches: Operators

By 1878, Bell had developed a system in which a local wire pair would come into a central location called an **exchange.** The exchange was a location that contained a **switchboard,** which was a collection of connectors and connector wires, that allowed a human controller, called an **operator,** to

FIGURE 5-1 Toggle switchboard, circa 1930s.

manually switch calls from caller to receiver. A manual toggle switchboard is shown in figure 5-1. Today, the term *central office* is commonly used to refer to an exchange although some still use the term *central exchange*.

A *switchboard* is a connection device that can consist of a series of plugs and jacks or toggle switches like those shown in figure 5-1. A caller would pick up the handset and turn a generator handle, which would cause a bell to ring on the switchboard. A human *operator* would answer the caller request, get the receiver party name or number to be called from the caller, and switch the connection using tip-and-ring plugs or toggle switches on the switchboard. This type of system is shown in schematic form in figure 5-2.

FIGURE 5-2 Schematic symbol for an exchange.

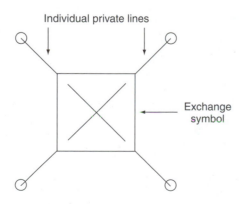

Individual private lines

Exchange symbol

FIGURE 5-3
Schematic symbol for
an exchange with
private and party
lines attached.

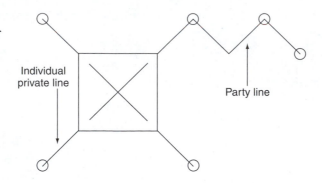

Initially, private single residence lines were considered a luxury and were restricted, because of cost, to wealthy residences and businesses. Many residences were connected with what is called a **party line.** A party line is a wire pair that is shared by two or more customers. A single wire pair is truly shared; if one customer wants to make a call and another is using the line, the customer wanting to use the line has to wait. This type of system is shown in schematic form in figure 5-3. This system worked great in small local environments where one operator could easily handle a relatively small number of calls. As more lines were run and more customers subscribed, the exchanges were geographically broken up and additional operators were used, as illustrated in figure 5-4.

FIGURE 5-4 Large
exchange broken up
into two.

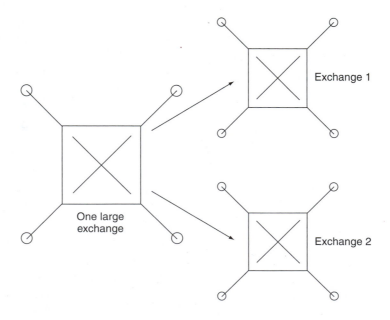

FIGURE 5-5
Trunked local
exchanges.

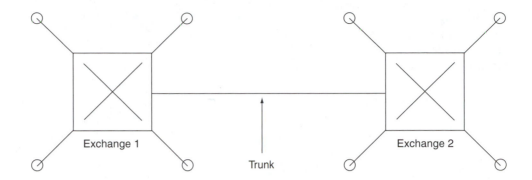

Trunking

Looking at figure 5-4 we can now see another problem. Originally, all customers on the single large exchange could be interconnected through the single exchange. When the exchange was split into two, some original customers went to exchange 1 and some went to exchange 2. What if a customer on exchange 1 wanted to talk to a customer on exchange 2? The solution was to attach the two smaller exchanges with high-capacity exchange-to-exchange transmission lines called **trunks.** An example of trunked local exchanges is shown in figure 5-5.

The trunk consisted of bundled wire pairs connecting the two exchanges. For example, when the operator on exchange 1 received a call from an exchange 1 party requesting connection to a party on exchange 2, the exchange 1 operator contacted the exchange 2 operator. The request was made, and the dedicated circuit wire pair was built by the two operators in the different exchanges connecting the two customers. The trunks had a finite number of pairs available for interexchange calls; and, if all pairs were being used, the caller would have to wait until a trunk pair became available.

Tandem Offices

The network continued to grow, and the need for centralization became apparent. **Tandem offices,** as illustrated in figure 5-6, are facilities that have been designed to concentrate trunk switching and serve an area approximately the size of a small town. Using the tandem office concept, all trunk switching is handled in the tandem office. This made it easier to add trunk wire pairs between exchanges as the network continued to grow. Typically, a tandem office would serve an area the size of a small town. All tandem office interconnections are considered to be local and not long-distance or toll calls. Customers pay a single flat fee to the local exchange carrier (LEC) in their town for use of the tandem office network.

FIGURE 5-6
Tandem office
connecting local
exchanges.

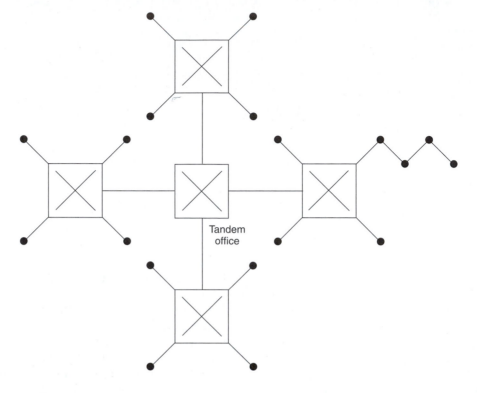

Toll Centers

As the network continued to grow, telephone companies began to intercon-nect tandem offices to provide long-distance calling to their customers. This interconnection was done using **toll offices** or **toll centers,** with the in-dividual call centers connected by lines called **toll-connecting trunks,** as illustrated in figure 5-7.

FIGURE 5-7 Toll-connected tandem offices.

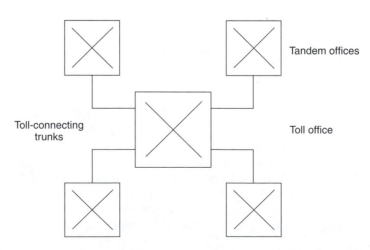

FIGURE 5-8 Toll-trunked toll centers.

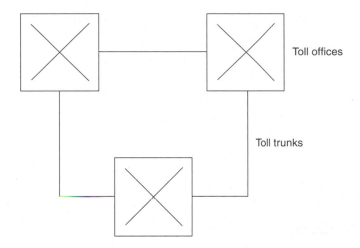

Toll offices

Toll trunks

Toll offices or toll centers are larger facilities than tandem offices with higher-capacity connecting lines. This construction, as illustrated in figure 5-8, resulted in a network similar to the one we now use today, allowing us to pick up a phone in our home or business and easily place a call to anywhere in the United States or most countries in the world.

5.2 Switching Emerges

That the initial telephone network was built and switched using manual operators is hard to believe. A caller would pick up a handset and signal the operator. The operator would pick up the call, request the number being called, and make the connection or pass the call on to another operator if required. Switchboards included lights indicating the status of each line because, once a call was complete, the operator had to pull the plugs and disconnect the circuit, freeing up lines for the next call. Manual switchboard construction was an art form in itself, with the goal being to place as many switch jacks as possible into an average arm's length of reach.

Legacy Switching Hierarchy

Earlier switches were put into a hierarchical classification structure. Even though today's modern long-distance network is considered a nonhierarchical dynamic system with all switching centers equal, the terminology of the legacy system is still commonly used today.

FIGURE 5-9
IntraLATA calls.

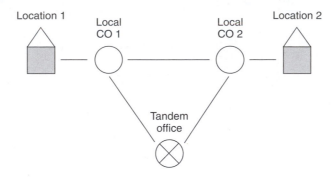

IntraLATA Calls

The exchange to which the local loop is connected is also called a *class 5 office*. We understand that the switch in the exchange or class 5 office determines the best route to where the call needs to go based on the number dialed. **Local access and transport area (LATA),** in the United States, refers to a geographic area covered by one or more local-exchange carriers (LECs). An **intraLATA call,** as illustrated in figure 5-9, is defined as a call between two LECs within a LATA. For example, calling from Cambridge, Massachusetts, to Boston, Massachusetts, is considered intraLATA. The call is handled by the LEC (Verizon) and may or may not be long distance. An intraLATA call may or may not include a tandem office in the connection. A tandem office connection will be used only when necessary.

InterLATA Calls

An **interLATA call** is defined as a call between a carrier in one LATA to a carrier in another LATA. InterLATA is long-distance service, as illustrated in figure 5-10. InterLATA calls are received by the local CO, the switch determines where the call needs to go, and the call is forwarded on to the **interexchange carrier (IXC)** class 4 toll center or **point of presence POP.** The point of presence defines the PIC or long-distance carrier that has been selected by the customer (e.g., AT&T, MCI, or Sprint).

Notice the redundancy built into the switched network in figure 5-10. The shortest route is always the cheapest route, but there are times when lines are busy or there are technical difficulties. When connections could not be directly made, class 4 centers passed on to class 3 centers, which passed to class 2 sectional centers, which passed to class 1 regional centers. The goal was to keep all connections as low on the list as possible because the higher a call went, the more expensive the connection got. AT&T originally built this model and even though, technically, it is not used in today's nonhierarchical long-distance system, much of the terminology is still in use today.

FIGURE 5-10
Legacy center-connection hierarchy.

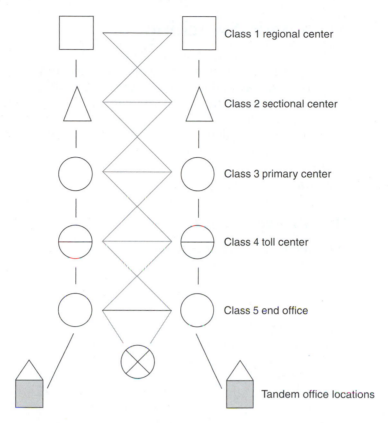

Class 1 regional center

Class 2 sectional center

Class 3 primary center

Class 4 toll center

Class 5 end office

Tandem office locations

Private-Branch Exchanges (PBXs)

Private-branch exchanges (PBXs) are exchanges used to switch calls within a business. The original PBXs were basically manual operator switchboards. Calls would come into a business, the switchboard operator would answer and ask callers to whom they would like to speak, and then physically make the proper connection using the same tip-and-ring plugs and jacks the telephone operators used.

In 1963, PBX computers were introduced by the Bell System and were used to handle the internal call switching. The Bell System installed and maintained all business PBXs in the United States until 1968, when the FCC ruled that outside vendors could come into a business and connect their equipment to Bell lines as long as they followed rule 68 guidelines, which specified all electrical and interface requirements for attachment to telephone lines.

Today's PBX topology resembles more of a computer network than a voice network. Typical PBXs consist of a digital switch and are true multimedia communications devices that integrate voice and data services, allowing the individual user the ability to manage all types of communications—voice mail, E-mail and faxes—right from a computer desktop.

5.3 Automatic Switching

As the network continued to grow, the demand for operators quickly began to outstrip the available supply. Manual circuit switching was also very expensive because of its labor intensity. In 1892, a Kansas City undertaker named Almon B. Strowger was upset because the wife of the competing funeral home director in Kansas City was also the switchboard operator for Kansas City. Calls would come in for a funeral home and would be connected by the operator to Strowger's competition. As a result, Strowger invented the first automatic switch.

The Strowger Switch

The **Strowger switch** is an electromechanical switching system set up in an array or *frame* of 10 rows and 10 columns, as shown in figure 5-11. Looking at the Strowger-switch frame in the figure, we can see that one frame has 100 individual switch positions and can handle the equivalent of 100 connections.

The switch itself is cylindrical, as shown by the Strowger-switch contact bank in figure 5-12. A series of relays and electromagnets forms the switch array, and, when numbers are dialed, pulses close the relays and activate the magnets causing a switch contact or *wiper* to move across the array to the proper switch connection. Strowger's switch was invented in 1891, 4 years before the telephone dialer was invented in 1895. To eliminate the need for the operator, Strowger had to determine a way to generate the pulses to move the relays. Strowger developed a system using three push buttons—one for

FIGURE 5-11
Strowger-switch frame.

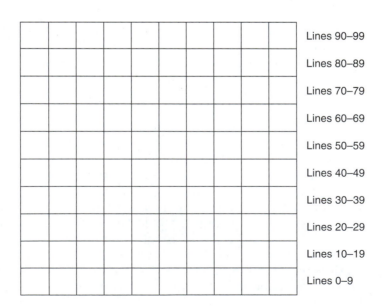

Lines 90–99

Lines 80–89

Lines 70–79

Lines 60–69

Lines 50–59

Lines 40–49

Lines 30–39

Lines 20–29

Lines 10–19

Lines 0–9

FIGURE 5-12
Strowger-switch
core.

hundreds, one for tens, and one for singles. If users wanted to call the number *258*, they would push the hundreds button 2 times, the tens button 5 times, and the singles button 8 times.

Strowger installed his first switch in his hometown of La Porte, Indiana, in 1892 with limited success. The Bell System companies were happy with manual switching at the time and not interested in his switch. In 1896, Strowger sold his patents to the group with which he had been working and then sold his share of the company to the same group for $10,000 in 1898. Strowger died in 1902. In 1916, the Bell System realized that the era of the manual operator was over and bought these same patents from Strowger's original group for $2.5 million. In 1919, Bell introduced dial telephones in Norfolk, Virginia, and the Strowger switch went into wide use across the world.

The Bell System preferred the term **step-by-step switch** instead of Strowger switch. A typical Bell step-by-step switch system installed in a CO was an array of Strowger frames that could handle 10,000 individual telephone lines. Walking into a busy CO was quite an experience as calls were being electromechanically switched. You can listen to a Strowger switch stepping at: http://www.light-straw.co.uk/ate/ar1/step1.wav

The Strowger switches required high maintenance, but they were very reliable. Their use continued in the United States until the late 1980s when computer modem prices dropped and data transmission on the voice network started to become popular. The electromechanical switches were electrically noisy and were quickly replaced by electronic switches. Strowger switches are still used in other parts of the world.

FIGURE 5-13
Crossbar-switch
array.

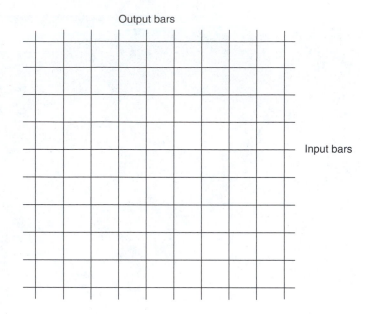

Output bars

Input bars

The Crossbar Switch

The Bell System soon realized the labor intensity of maintaining Strowger switches and developed the **crossbar switch,** an electromechanical switching system with fewer moving parts than the Strowger. In 1938, the Bell System began installation of *number 1 crossbar* switches in large cities. The crossbar switch also uses a matrix of individual switch connections, as illustrated in figure 5-13.

In the crossbar-switch array, the input and output lines are not actually touching until contact is made. Contact in the switch is controlled by magnets; when an input bar is touching an output bar at a crosspoint, a switch connection is made, as illustrated in figure 5-14. The crosspoint contacts actually latch and will hold contact position until the call is completed and the contact is disconnected.

Electronic Switches

Crossbar switches are very reliable and, even though they are also electromechanical, do not require the amount of constant maintenance that a

FIGURE 5-14
Crossbar crosspoint,
expanded view.

No crossbar contact

Crossbar contact

Strowger switch does. As late as 2000, crossbar switches remained very popular in other countries and were still in use in some parts of the United States. However, most of the United States has switched or is rapidly switching over to electronic switching systems.

The Electronic Switching System (ESS)

The first popular electronic switch was the *Number 1 Electronic Switching System (No. 1 ESS)* installed by AT&T in 1965. **Electronic switching systems (ESS)** are computerized switches with very few moving parts. The *No. 1 ESS* switch can automatically handle up to 65,000 individual lines and up to 25,000 calls per hour. In 1976, AT&T installed the first *No. 1A ESS*. This switch can automatically handle up to 128,000 individual lines and up to 110,000 calls per hour.

Remember that a physical connection must still be made and a fair amount of current must flow through this physical connection to complete a telephone circuit. Both Number 1 ESS switches use an array of contacts similar to the crossbar system but do not use actual wires or bars to make physical connections. The No. 1 ESS uses a *ferreed switch,* and the No. 1A ESS uses a more advanced *remreed switch* to make the physical contact required to complete a switch connection. Both the *ferreed switch* and the *remreed switch* are **reed switches** made of a magnetic alloy called **remendur** and are sealed in a glass tube. The remendur alloy allows them to operate in two states, on or off. Computer-controlled pulse distributors in the switch generate current that flows through an electromagnetic coil surrounding the individual reed switches. The remendur alloy can be made magnetic by applying electric current through the coil in one direction and nonmagnetic by applying electric current through the coil in the opposite direction. The current direction through the coil determines the polarity of the magnetic field produced by current flow in the wire coil, and the switch will open and close as illustrated in figure 5-15. The contact shapes resemble musical reeds, hence the names *ferreed* and *remreed.* Current is not required to maintain the state of the switch, instantaneous current, and the corresponding polarized magnetic field; current is only required to open or close the switch.

The No. 1 and No. 1A ESS switches are fundamentally computers that control the opening and closing of these miniature reed switches. The first ones developed in the 1960s and 1970s used *ferrite* or *core* computer memory, but most have been upgraded now to use *random access memory.* The switches run line scanners that monitor incoming lines, reed switch states ,and all incoming and outgoing traffic, setting up call switching when a call needs to

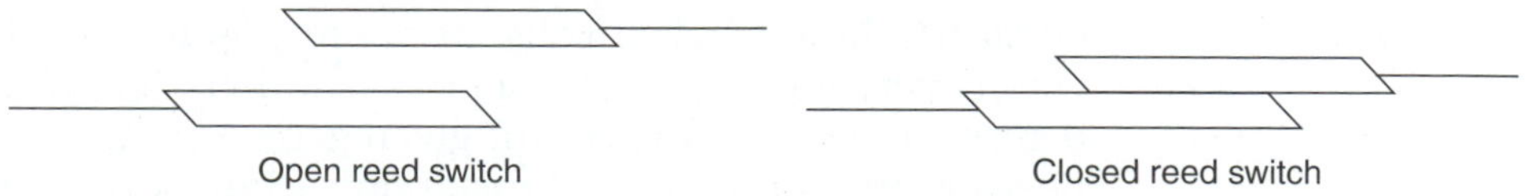

FIGURE 5-15 Reed switch, open and closed.

Open reed switch   Closed reed switch

TABLE 5-1 Some common ESS switches.

Switch spec	Use	Lines	Calls per hour	Release date
No. 4 ESS	Toll/tandem exchange	53,760	550,000	1976
No. 5 ESS	Local exchanges	100,000	650,000	1982

be completed and removing it when a call is finished. Two other ESS switches are in common use today, as specified in table 5-1.

The Digital Multiplex System (DMS)

In addition to the ESS system developed by AT&T, Northern Telecom developed the *digital multiplex system (DMS)* series of switches. The DMS is a family of time division multiplexed digital switching systems for local, tandem, and toll applications. We can briefly look at a few of the more popular models installed in telephone company central offices.

DMS10-E

The Northern Telecom DMS-l0E digital switching system is a small digital switch that can be expanded to 8000 subscriber lines. The switch offers all of the currently popular custom calling features.

DMS-100

The DMS-100 digital switching system is a large local CO digital switching system that can be expanded to 100,000 subscriber lines.

DMS-200

The DMS-200 digital switching system is a large toll/tandem digital switching system that can handle up to 60,000 trunk circuits.

5.4 Traffic and Blocking

We now have an idea of how a switch works and understand that different switches have the capacity to handle different finite amounts of traffic volume. If traffic gets too heavy on any one switch, **blocking** of calls will occur. A blocked call is a call that is not allowed to complete. Maybe you have experienced the *fast busy* signal we described in the telephone set discussion. These blocked calls are more popular on days like Mother's Day, when everyone is trying to call their mothers! **Traffic** engineering is used to determine switch capacity for given geographical areas, and the traffic engineering procedure can be broken down into several steps.

Step 1: Determine Busy Hour Traffic

Traffic engineers specify a time period and monitor call traffic over that period. Over this period, several things are measured.

Busy Hour

Busy hour is defined as the hour with the most traffic in the specified time period. During this hour, some calls may be blocked if switch capacity is exceeded.

Traffic Density

Traffic density is defined as a maximum traffic value for the defined busy hour and is expressed in **century call second (CCS)** units. To calculate a traffic density value in CCSs, busy hour traffic is monitored and calls are counted and averaged. The call count is then multiplied by the average time for each call, and this number is then divided by 100.

CCS = (call count)(average time per call)/100

Consider example 5.1.

●—EXAMPLE 5.1

●—Problem
Traffic engineers are monitoring a local CO switch. At the busy hour, they measure 9700 calls with an average time of 145 seconds per call. Calculate the traffic density.

●—Solution

CCS = (call count)(avg time/call in seconds)/100
    = (9700 calls)(145 seconds/call)/100 = 14,065 CCS

The information in example 5.1 is important to traffic engineers because switches and PBXs come with maximum CCS ratings that cannot be exceeded or calls will be dropped.

Grade of Service

The **grade of service** is the probability that a call will be blocked at a given CCS rate. For example, a grade of service of .02 at 14,000 CCS means 2 calls per 100 will be blocked at the 14,000 CCS rate.

Erlang's

A. K. Erlang of Denmark was the first person to study the traffic problem on voice telephone networks. In 1917, he published a paper titled *Solution of*

Some Problems in the Theory of Probabilities of Significance in Automatic Telephone Exchanges. In this paper, he presented two calculations for telephone traffic loss and waiting time and, by the end of World War II, his name was used worldwide to denote the unit of telephone traffic.

The **Erlang** unit is equal to 36 CCS. The single unit is based on 1 call lasting 1 hour (3600 seconds). Recall that

CCS = (call count)(average time per call)/100

If the call count is 1 and the average time per call is 3600 seconds, then:

CCS = (1 call)(3600 seconds/call) = 36 CCS

Consider example 5.2.

•—EXAMPLE 5.2

•—Problem
For example 5.1, we calculated a value of 14,065 CCS. Convert this calculation to Erlangs.

•—Solution

(14,065 CCS)/(36 CCS) = 390.69 Erlangs

Load Factor

The **load factor** (*C*) indicates the average number of calls that can be made at the same time during the busy hour. We can calculate the load factor as follows:

C = (number of calls)(average time per call in seconds)/3600

•—EXAMPLE 5.3

•—Problem
Go back to example 5.1. At the busy hour, engineers measure 9700 calls with an average time of 145 seconds per call. Calculate the load factor.

•—Solution

C = (9700 calls)(145 seconds per call)/3600 seconds per hour = 390

The small switch in this example can carry 390 calls per hour.

Two models for blocked call handling are used by engineers. The *Poisson model* assumes that any single blocked call would be tried again in a short period of time, and the *Erlang-B model* assumes that any single blocked call would be dropped and not tried again. The different models are used for different traffic engineering scenarios. Data for both models are calculated and tabulated in tables; engineers, after calculating the CCS and expected grade of service, use the tables to determine switch capacities. Many people who have added second voice lines for data in their homes dial up with their modems and stay almost permanently connected. This can cause major calculation problems for the traffic engineers and is one of the reasons telephone companies are moving to provide alternative nonswitched data services like *asymmetric digital subscriber line (ADSL)*.

Summary

1. The first switching systems were called *exchanges* and were run by human *operators* using *switchboards.*

2. Most early residential connections were *party lines* with a wire pair being shared by two or more customers.

3. As the phone system grew, a *hierarchical system* was built connecting network to network.

4. A *local access and transport area (LATA)* is a U.S. term that defines a geographical area covered by one or more local telephone companies. Calls can be *intraLATA* or *interLATA*.

5. InterLATA calls are considered *long distance* and are switched to the customer's selected long-distance carrier *point of presence (POP)*.

6. *Private-branch exchanges (PBXs)* switch calls internally within a business.

7. The *Strowger switch* was the first automatic switch developed in 1892 and went into wide use in 1919 when the Bell System introduced dial telephones.

8. In 1938, the Bell System developed the *crossbar switch*, which had fewer moving parts than the Strowger and was easier to maintain.

9. The first *electronic switching system (ESS)* was installed by AT&T in 1965.

10. Electronic switches use *reed switches* to make the physical contact required to complete a call.

11. Switches have a finite number of physical connections, and, when operating at full capacity, calls can be *blocked* or not allowed to complete.

12. Traffic engineers have defined and monitor different values to manage traffic. These values include *busy hour, traffic density, grade of service,* and *load factor.*

13. The *Erlang unit* is used to denote telephone traffic.

Review Questions

Section 5.1

1. The first switching system developed was called a/an _____.
   a. breadboard
   b. switchboard
   c. reedboard
   d. exchange

2. When more than one customer share a single wire pair, the connection is most commonly referred to as a _____ line.
   a. group
   b. local
   c. shared
   d. party

3. _____ transmission lines are used to connect telephone company exchanges.
   a. Switch
   b. Single-wire
   c. Trunk
   d. operator

4. Trunk switching is typically handled at a/an _____ office.
   a. operator
   b. exchange
   c. toll center
   d. tandem

5. _____ are used to interconnect tandem offices and provide long distance calling.
   a. Exchanges
   b. Toll centers
   c. RBOCs
   d. LECs

Section 5.2

6. A connection between two LECs within a LATA is referred to as _____.
   a. long distance
   b. local
   c. intraLATA
   d. interLATA

7. A connection between a carrier in one LATA to a carrier in another LATA is referred to as _____.
   a. PBX
   b. local
   c. intraLATA
   d. interLATA

8. IntraLATA calls are always considered long distance. True/False

9. InterLATA calls are always considered long distance. True/False

10. _____ were developed to switch calls internally within a business.
    a. Exchanges
    b. Toll centers
    c. LECs
    d. PBXs

Section 5.3

11. The _____ switch was the first automatic switch developed.
    a. No. 1 ESS
    b. Strowger
    c. step-by-step
    d. crossbar

12. Mechanical switches are no longer common in the United States but are common in other parts of the world. True/False

13. The _____ electromechanical switch had fewer moving parts than earlier switches and was first installed in 1938.
    a. No. 1 ESS
    b. Strowger
    c. step-by-step
    d. crossbar

14. The first popular electronic switch was the _____.
    a. No. 1 ESS
    b. Strowger
    c. step-by-step
    d. crossbar

15. Electronic switches use reeds made from _____.
    a. nickle
    b. gold
    c. iron
    d. remendur

Section 5.4

16. When a switch capacity is full, calls coming into that switch are said to be _____.
    a. open
    b. blocked
    c. shorted
    d. shunted

17. _____ is defined as the hour with the most traffic in a specified time period.
    a. Traffic density
    b. Grade of service
    c. Busy hour
    d. Load factor

18. _____ is the probability that a call will be blocked at a given CSS rate.
    a. Traffic density
    b. Grade of service
    c. Busy hour
    d. Load factor

19. _____ are the units used to denote telephone traffic.
    a. Watts
    b. Erlangs
    c. Ohms
    d. Amperes

20. _____ indicates the average number of calls that can be made at the same time during the busy hour.
    a. Traffic density
    b. Grade of service
    c. CSS
    d. Load factor

21. The _____ model assumes that any blocked call will be tried again in a short period of time.
    a. CSS
    b. Erlang-B
    c. Poisson
    d. Bell

22. One Erlang is equal to _____ CSS.
    a. 2
    b. 12
    c. 24
    d. 36

23. The _____ model assumes that any single blocked call will be dropped and not tried again.
   a. CSS
   b. Erlang-B
   c. Poisson
   d. Bell

24. Switch traffic density is indicated using _____ units.
   a. CSS
   b. Erlang
   c. cycles per second
   d. hertz

Discussion Questions

Section 5.1

1. What is a party line? Are party lines still used in the United States?

2. The term *trunk* is a common one with regard to the telephone system. Describe three different types of trunks used in the telephone network. Be sure to use sketches of examples of each in your answers.

Section 5.2

3. Explain the difference between intraLATA and interLATA calls.

4. What is a PBX? How are PBXs used? Give an example of a PBX you have seen.

5. What type of PBX is used on your campus? Be sure to list the manufacturer and model number.

Section 5.3

6. How was the first electromechanical switch invented? Who invented it? How long did it take for this first electromechanical switch to be widely used? What had to happen before the switch went into wide use?

7. What are the major differences between the Strowger switch and the step-by-step switch?

8. In the United States, approximately when was electromechanical switching replaced by computerized switching? What caused this change?

9. List two advantages of the crossbar switch over earlier electromechanical switches.

10. What was the first popular electronic switch used in the United States? What was the call capacity of this switch?

11. How is physical contact made in an electronic switch to complete a circuit? Give details.

Section 5.4

12. Describe what has happened when you pick up a telephone, dial a number, and get a fast busy signal.

13. Traffic engineers use the term *busy hour*. Define this term.

14. A switch is being monitored by traffic engineers. At busy hour, they are measuring 4500 calls that last an average of 123 seconds each. Calculate the traffic density.

15. A switch is being monitored by traffic engineers. At busy hour, they are measuring 5550 calls that last an average of 200 seconds each. Calculate the traffic density.

16. Traffic engineers obtain the following information when monitoring a switch at a 9000 CCS rate: 18 calls out of 600 are blocked. Calculate the grade of service.

17. Traffic engineers calculate a traffic density value of 12,657 CCS. Calculate the traffic density in Erlangs.

18. Convert your traffic density value obtained in problem 14 to Erlangs.

19. Convert your traffic density value obtained in problem 15 to Erlangs.

20. A switch is being monitored by traffic engineers. At busy hour, they are measuring 6500 calls that last an average of 432 seconds each. Calculate the load factor.

21. A switch is being monitored by traffic engineers. At busy hour, they are measuring 9000 calls that last an average of 127 seconds each. Calculate the load factor.

22. Describe the difference between the Poisson and the Erlang-B models used by traffic engineers.

23. How does the addition of an added voice line for PC/modem use affect traffic calculations? Be specific.

24. xDSL and other broadband services are "always on" and are considered switchless technologies because they bypass telephone company switches. If the switches are bypassed, do traffic engineers need to consider these types of connections for switch calculations? Why or why not?

Signaling

Objectives   Upon completion of this chapter, the student should be able to:

- Understand the importance of signaling.
- Describe how calls move from one point to another.
- Discuss in-band signaling.
- Comprehend why out-of-band signaling is now used.
- Describe the out-of-band signaling process.
- Explain SS7 network details.
- Understand how ISDN networks function and integrate into the existing switched network.
- Describe ISDN signaling.
- Discuss broadband ISDN.

Key Terms

B-channel
basic-rate interface (BRI)
broadband
broadband ISDN (B-ISDN)
D-channel
hacker

in-band signaling
integrated services digital network (ISDN)
International Telephone and Telegraph Consultative Committee (CCITT)

multifrequency (MF) tones
non-facility-associated signaling (NFAS)
out-of-band signaling
phreak
primary-rate interface (PRI)

signal control point
(SCP)

signal switching point
(SSP)

signal transfer point
(STP)

single-frequency (SF)
tones

Introduction

Signaling on the voice network can be considered as anything that is not voice that is passed through the network. For example, picking up a telephone receiver generates a current on the local loop and signals the attached CO switch that a line has been picked up. When you dial numbers on your home telephone, you are sending signals to a switch attached to the copper wire pair that comes into your home. Both of these cases would be considered *local loop* or *local signaling.* When making long-distance calls, in addition to local loop signaling, *long-distance signaling* is required between your local switch and the local switch of the person you are trying to call. Have you ever made a long-distance call and had a busy signal returned to you? If so, you have experienced long-distance signaling! Signaling on the telephone network can be either *analog* or *digital* and either *in-band* or *out-of-band.*

6.1 How Calls Move from One Point to Another

Before discussing call signaling in more detail, let us look at how calls are directed from one point to another. In chapter 3, we learned that the local loop is connected to what is called an *exchange* or *class 5 office.* The switch in the class 5 office determines the best route to where the call needs to go based on the number dialed. A connection between two local exchanges within the LATA is referred to as *intraLATA.*

A connection between a carrier in one LATA to a carrier in another LATA is referred to as *interLATA* or *interexchange.* InterLATA is long-distance service. InterLATA calls are received by the local CO, and the switch determines where the call needs to go. The call is forwarded to the long-distance carrier point of presence (POP) that has been selected by the customer.

In-Band Signaling

In-band signaling uses signals that are sent over the same channels as voice signals. Chapter 3 explained how picking up a telephone receiver signals the attached CO. The chapter also explained how numbers dialed on a telephone are transmitted to the CO switch for interpretation. A telephone connected to the local loop uses in-band analog signaling. One wire pair connects the telephone to the CO switch, and all signals must travel over this single wire pair. Up until 1975, the long-distance network also used in-band signaling.

In the mid-1950s AT&T decided to operate its long-distance switching system using 12 electronically generated combinations of 12 master signal tones. The long-distance network used **single-frequency (SF)** and **multi-frequency (MF) tones** for these signals. Single-frequency (SF) tones are tones that are made up of only one frequency, while multifrequency (MF) tones are those that are made up of more than one frequency.

These tones are different from the dual-tone-multifrequency (DTMF) tones discussed in chapter 3; in fact, these tones were selected before DTMF was invented. These long-distance tones were kept "secret" until the 1960s when an AT&T engineer wrote an article on switching theory for a technical journal. In the article were the actual frequencies used; these frequencies are shown in table 6-1. The really critical frequency here was the 2600-hertz IDLE single frequency that was used to identify an idle condition on a trunk line. If a trunk line was available, it would have a 2600-hertz tone on it and it could be picked up and used by the switch to carry a long-distance call.

TABLE 6-1 In-band long-distance frequencies.

Number/signal	Frequency
IDLE	2600 Hz
START	1500 Hz and 1700 Hz, combination
KP	1100 Hz and 1700 Hz, combination
0	1300 Hz and 1500 Hz, combination
9	1100 Hz and 1500 Hz, combination
8	900 Hz and 1500 Hz, combination
7	700 Hz and 1500 Hz, combination
6	1100 Hz and 1300 Hz, combination
5	900 Hz and 1300 Hz, combination
4	700 Hz and 1300 Hz, combination
3	900 Hz and 1100 Hz, combination
2	700 Hz and 1100 Hz, combination
1	700 Hz and 900 Hz, combination

Telephone **hackers,** also called phone **phreaks,** are people who seek vulnerabilities in the telephone system and attempt to take advantage of them. Phone phreaks originally used cheap home entertainment organs to produce these same frequencies and record them. Once they had the correct frequencies recorded, they would then use a second recorder to put together series of numbers. Once the numbers were recorded, the recorder speaker was held up to the telephone transmitter and the tones were played and used to obtain free long-distance telephone calls. Many phone phreaks built electronic tone generators called *blue boxes* that had push buttons to generate the different frequencies. One of the original phone phreaks was a man named Joe Engressia, who did not even use a blue box; he could *whistle* into the phone at the proper frequencies! As long as the recorded frequencies were within 30 hertz of the phone company's tones, the telephone switch equipment would "think" it was hearing tones generated by other telephone company equipment.

These blue box tone generators were used to make free long-distance calls in the following way:

1. A phreak would usually go to a pay phone and dial any valid toll-free 800 number. Dialing an 800 number tells the telephone switch that the call will not be billable.

2. When the 800 number started to ring, the phreak would hold the blue box speaker to the receiver and generate a 2600-hertz idle tone that the switch would "hear."

3. When the 2600-hertz tone stopped, the switch would assume the line was once again being used so it would "listen" for a new series of tones to find out where the call should be sent.

4. The phreak would then beep out the 10 digits of the long-distance number on the blue box and make a free call!

One of the most famous early phone phreaks was a man who went by the name of Captain Crunch and was named after the breakfast cereal. At one time, the cereal included a toy whistle that blew a perfect 2600-hertz idle tone! The story goes that the man Captain Crunch was in the Air Force and was transferred to Europe. When his friends would call him, he would blow his Captain Crunch whistle into his phone and make the calls free for his callers.

Out-of-Band Signaling

Once released, these in-band frequencies were very easy to duplicate and AT&T was losing lots of money as blue box popularity grew. As a result ,AT&T developed an **out-of-band signaling** system that replaced the in-band, long-distance MF signaling. **Out-of-band** signals are signals that are carried on a separate channel from the voice channel. This made the blue box tone generators obsolete. This out-of-band signaling system is referred to as the *Signaling System 7 (SS7) network.*

The Signaling System 7 (SS7) network is a completely separate network from the one that carries voice; as a result, signaling occurs at a much faster rate. It works in the following way:

1. A caller picks up a handset and dials a long-distance number.

2. The switch in the local class 5 CO determines the best route to where the call needs to go based on the number dialed.

3. If the number is long distance, the call is forwarded to the interexchange carrier (IXC) class 4 toll center that has been identified as the customer's preferred interexchange carrier.

4. The IXC switch sends a request over the SS7 network that will test the called number and also find a route for the call to take.

5. If the called number is busy, the first or originating IXC switch sends a busy signal to the caller. This signal is not sent over the long-distance voice network; SS7 signals back to the originating IXC switch, and the busy tone is generated there.

6. If the number is not busy, the SS7 network builds and reserves a circuit from the caller class 5 CO to the receiver class 5 CO over the long-distance voice network.

7. The receiver class 5 CO rings the number being called.

8. When the receiver answers the call, the reserved long-distance voice circuit is activated and the long-distance voice network is in use until the call is completed.

9. If the receiver does not answer the call and the caller hangs up, the SS7 network will release the reserved circuit connections without actually using the long-distance voice network.

The SS7 network does not carry voice; it is a separate network designed for signaling data only and, as a result, is much faster than the old in-band signaling method. SS7 also allows signaling throughout an ongoing telephone call. With in-band signaling, once the call was set up, switches could not communicate without a direct trunk connection. SS7 is also required for the custom local area signaling services (CLASS), like caller ID and call blocking, discussed in chapter 3. Call establishment, billing, routing, and the exchange of information between switches is handled by SS7. SS7 is also required for ISDN D-channel use.

SS7 Details

You can probably see now that the SS7 network could be called an intelligent data network. This network is made up of signal switching points, signal transfer points, and signal control points, as follows:

Signal switching points (SSPs) are the SS7 switches at both the caller and receiver ends. They are the class 5 CO switches that are SS7 enabled. These switches can originate, terminate, or switch telephone calls.

Signal transfer points (STPs) are the SS7-enabled switches that handle the long-distance path of SS7 signal traffic. These switches receive and route incoming signals to the required destination.

Signal control points (SCPs) are databases that work with the STPs. These databases provide information about advanced call-processing capabilities that are provided and used by the network.

Usually, STPs and SCPs are installed in pairs to build redundancy into the SS7 network. However, they do not have to be installed in pairs. Two identically paired STPs are referred to as *mated-pair STPs,* and two identically paired SCPs are referred to as *mated-pair SCPs*. The universal schematic symbols for these units are shown in figure 6-1. The way that these units connect a telephone into the rest of the network is shown in figure 6-2. The system redundancy is easy to see here. The single telephone is attached to the SSP-enabled switch in the local CO. The single SSP-enabled switch in the local CO is attached to two separate mated-pair STPs. Attached to the two mated-pair

FIGURE 6-1 SS7 unit schematic symbols.

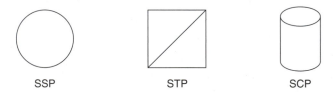

FIGURE 6-2 SS7 network connections.

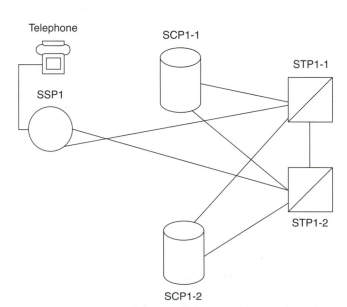

STPs are two mated-pair SCPs. Typically, the two STPs are not in the same location and the two SCPs are not in the same location. In the diagram, SCP1-1 and SCP1-2 are mirror-image databases that control STP1-1 and STP1-2. The interconnection lines are referred to as *signal links,* and signals move on these links at either 56 kilobits per second or 64 kilobits per second. With the interconnection shown in the figure, if one SCP and one STP go down at the same time, the connected telephone can still make long-distance calls.

Redundancy is also built into the long-distance portion of the SS7 network. The way the STPs in two interconnected SS7 networks may be connected is shown in figure 6-3. Each thick line in the figure represents an interconnection between STPs; once again, the redundancy can be seen. Up to four of the signal links can go down and the two networks will still be able to communicate with each other, as illustrated in figure 6-4. In the figure, only the STP1-1 and STP2-1 signal link connection between SS7 networks 1 and 2 remains. Even though four of the five signal trunks are not operational, the two SS7 networks can communicate and call paths can be set up.

FIGURE 6-3
Interconnected SS7 network STPs.

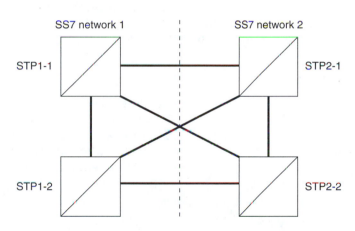

FIGURE 6-4
Interconnected SS7 network with down STPs.

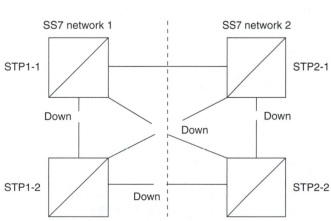

6.2 Integrated Services Digital Network

We have seen the evolution and now have a good understanding of the plain old telephone system (POTS) or public switched telephone network (PSTN) that was designed primarily for voice traffic. In most cases, only the local loop is still analog in signal format. The rest of the network works in digital format with the analog-to-digital and digital-to-analog conversions happening at the local central offices (COs). This is illustrated in figure 6-5.

In 1984, the **International Telephone and Telegraph Consultative Committee (CCITT),** now known as the previously discussed *International Telecommunications Union (ITU),* envisioned an all digital network and released initial guidelines on a worldwide standard for an all digital network. They called this standard the **integrated services digital network (ISDN).** The telephone companies initially felt that this standard was their network of the future, but development and installation have stalled for a number of reasons. Equipment is relatively expensive to install, and Northern Telecom (now Nortel Networks) and AT&T (now Lucent), who made the first ISDN switches, did not make the switches compatible with each other. Let us take a look at how ISDN works.

How ISDN Works

The integrated services digital network (ISDN) is an all-digital switched network that uses existing local loop lines to carry digital signals. This is illustrated in figure 6-6. These digital signals can be voice, data, fax, multimedia/video, and so on. Any signal that can be digitized can be transmitted via an ISDN provided the proper equipment is installed. In the local CO, an ISDN digital switch is required.

FIGURE 6-5 Plain old telephone system (POTS) block diagram.

FIGURE 6-6 ISDN system block diagram.

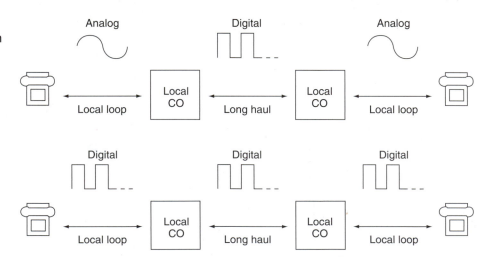

Using the PSTN's analog switches, we are limited to a single voice channel transmission per wire pair. The ISDN, in digital format, allows multiple channels to be transmitted over the local loop. The two ISDN access rates are **basic-rate interface (BRI)** and **primary-rate interface (PRI).** Each rate uses channels designated as **B-channels** (bearer channels) or **D-channels** (delta channels). In the ISDN, *B-channels* are used to transmit digital information and *D-channels* are used for signaling.

Basic-Rate Interface

The ISDN basic-rate interface (BRI) or ISDN basic access uses three channels for voice, data, and signaling and is commonly referred to as *2B+D* service. This 2B+D service is used primarily by home and small business users. Two 64-kilobits-per-second B-channels are used for voice and data transmission, and a single 16-kilobits-per-second D-channel is used for signaling. The D-channel handles all signaling that is done. This 2B+D service is illustrated in figure 6-7.

In the ISDN world, the *K* prefix in the abbreviation *Kbps* (kilobits per second) is a true metric prefix and refers to 1000 or 10^3, not the 1024 or 2^{10} that is used to designate things like computer memory and hard-drive capacity. The two 64-kilobits-per-second ISDN primary-access B-channels form a combined bandwidth of 128 kilobits per second:

(2 B-channels)(64 Kbps per channel)

United States telephone companies provide BRI customers with a connection called a *U-interface*. The U-interface is a wire pair (local loop) connecting the customer location to the ISDN switch. At the customer location, only one ISDN device can be connected to this two-wire pair because only one ISDN device can "talk" in bidirectional or full-duplex mode over two wires. A *network termination 1 (NT-1)*, also sometimes *incorrectly* referred to as an *ISDN modem,* is used to solve this problem. The NT-1 is used to convert

FIGURE 6-7 ISDN basic-rate interface (BRI).

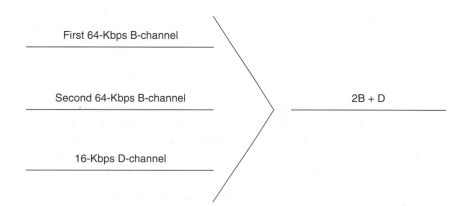

First 64-Kbps B-channel

Second 64-Kbps B-channel

16-Kbps D-channel

2B + D

FIGURE 6-8 ISDN NT-1 diagram.

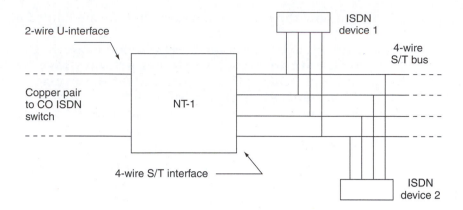

the two-wire U-interface into a four-wire S/T-interface. The S/T-interface is a four-wire interface; two of the S/T wires are used to transmit, and two are used to receive information. An S/T-interface bus cable can have up to seven ISDN devices attached to it, as shown in figure 6-8.

Note that ISDNs require ISDN telephones, fax machines, and other devices. Analog devices, like the telephones most of us have in our homes now, cannot be plugged into an ISDN; they will not work. Most ISDN devices being manufactured today come with two options. They can be purchased with a built-in NT-1 or U-interface, or they can be purchased with an S/T-interface that requires an external NT-1.

Primary-Rate Interface

The ISDN primary-rate interface (PRI) or primary access uses 24 channels for voice, data, and signaling and is commonly referred to as *23B+D* service because a total of 24 channels are used. This 23B+D service is used primarily by larger business users and is essentially the equivalent of a T-1 carrier circuit. Twenty-three, 64-kilobits-per-second B-channels are used for voice and data transmission, and a single 64-kilobits-per-second D-channel is used for signaling, as illustrated in figure 6-9. The primary-access D-channel also handles all signaling that is done. The 23, 64-kilobits-per-second ISDN primary-access B-channels form a combined bandwidth of 1.472 megabits per second:

(23 B-channels)(64 Kbps per channel)

which is approximately the bandwidth equivalent of a 1.544-megabits-per-second T-1 carrier service.

In Europe, primary rate is configured slightly differently. *European ISDN primary-rate interface (PRI)* or primary access is referred to as 30B+D service because a total of 32 channels are used. Thirty 64-kilobits-per-second channels are used for voice and data, and two 64-kilobits-per-second channels are used for signaling. Thirty B-channels produce a line rate of 1.920 megabits per second.

30 channels × 64 Kbps per channel = 1.920 Mbps

This bit rate is the approximate equivalent of an E-1 carrier circuit.

FIGURE 6-9 ISDN primary-rate interface.

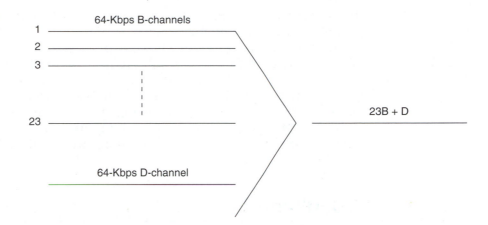

Non-facility associated signaling (NFAS) is a developed method that allows multiple PRI lines to be combined and controlled with one 64-kilobits-per-second D-channel. The ISDN system does this by aggregating B-channels into *H-channels*. H-channels are formed as shown in table 6-2.

TABLE 6-2 H-channel formation.

H-channel designation	Number of aggregated B-channels	Bit rate
H-0	6	384 Kbps
H-10	23	1472 Kbps
H-11	24	1536 Kbps
*H-12	30	1920 Kbps

*European system only

ISDN Signaling

The SS7 network is required for ISDN to work with the ISDN D-channel integrated directly into the SS7 network. Recall that signaling to a local-loop-connected telephone on the POTS network requires in-band signaling. An example of in-band signaling would be the 90-volt$_{rms}$ 20-hertz ringing signal that gets sent across the local loop on the single-channel wire pair. The ISDN, by keeping the signaling on its own channel, provides several advantages.

Out-of-band signaling used by the ISDN on the local loop allows signaling to occur throughout a communications session. For example, suppose that you are using your ISDN-modem-equipped computer to surf the Internet. With only one device requiring connection, the ISDN switch will provide both B-channels or 128 kilobits per second (2 × 64 Kbps) of bandwidth to the ISDN modem. If someone now calls your number, the D-channel will signal the ISDN telephone to ring and one of the B-channels will be dropped, allowing the voice session to occur simultaneously with the data session on each of the two separate B-channels.

Another advantage of the out-of-band signaling used by ISDN is the speed at which calls can be set up. We have all experienced the amount of time it takes to establish an analog modem dial-up connection on our personal computers. At times, call setup can take close to a minute. The ISDN call setup is very quick, typically taking anywhere from 1 to 4 seconds.

The ISDN signaling also includes things like caller ID and notification of whether the incoming call is a data call or a voice call.

6.3   Broadband ISDN

Both BRI and PRI ISDN are referred to as being *narrowband ISDN* because they are provided over limited-bandwidth copper wires. **Broadband ISDN,** or B-ISDN, is ISDN service provided in a **broadband** environment over fiber, using the high-bandwidth capabilities of fiber combined with asynchronous transfer mode (ATM) and synchronous optical network (SONET) transmission methods. Bit rates for B-ISDN can reach over 100 megabits per second.

Summary

1. *Signaling* on the voice network is considered any information transmitted that is not voice.

2. *In-band signals* are signals that are sent over the same lines as voice signals.

3. A telephone connected to the local loop uses *in-band analog signaling.*

4. Original telephone *hackers* were called phone *phreaks.*

5. *Blue box tone generators* were used with the early in-band signaling network to make free long-distance calls.

6. *Out-of-band signaling* was developed to prevent phone phreaks from stealing long-distance service.

7. The *Signaling System 7 (SS7)* network is an out-of-band network used only for signaling that is separate from the voice network.

8. The *Signaling System 7 (SS7)* network is a *redundant* system.

9. The *Integrated services digital network (ISDN)* is a worldwide standard for an all-digital network.

10. The *ISDN* uses existing local loop lines to carry digital signals.

11. The ISDN require *ISDN digital switches* in the local CO.

12. There are two basic ISDN rates, *basic rate interface (BRI)* and *primary-rate interface (PRI).*

13. Each ISDN uses channels designated as either *B-channels* (bearer channels) or *D-channels* (delta channels).

14. The ISDN B-channels are used to transmit *digital information.*

15. The ISDN D-channels are used for *signaling.*

16. The ISDN BRI is also referred to as *2B+D service* because two, 64-kilobits-per-second B-channels are used for data and one, 16-kilobits-per-second D-channel is used for signaling.

17. The ISDNs require *ISDN telephones* for voice use.

18. In the United States, ISDN PRI is also referred to as *23B+D* service because 23, 64-kilobits-per-second B-channels are used for data and one, 64-kilobits-per-second D-channel is used for signaling.

19. The *ISDN PRI service* is the equivalent of a T-1 carrier circuit service.

20. *Broadband ISDN* uses high-bandwidth fiber combined with asynchronous transfer mode (ATM) and synchronous optical network (SONET) transmission methods.

Review Questions

Section 6.1

1. A telephone call between two local exchanges within a LATA is referred to as _____.
   a. local loop
   b. intraLATA
   c. CLEC
   d. interLATA

2. A telephone call between two different LATAs is referred to as _____.
   a. local loop
   b. intraLATA
   c. CLEC
   d. interLATA

3. Long-distance calls are forwarded to a specific provider _____ (e.g., Sprint or MCI) that has been selected by the customer for long-distance service.
   a. LEC
   b. CLEC
   c. POP
   d. CO

4. In-band signals are signals that are mixed with voice signals on the same lines. True/False

5. The long-distance networks now use _____ signaling to operate long-distance switching systems.
   a. in-band
   b. out-of-band
   c. single-frequency (SF)
   d. multifrequency (MF)

6. The copper wire local loop uses _____ signaling to communicate with the connected switch.
   a. in-band
   b. out-of-band
   c. single-frequency (SF)
   d. multifrequency (MF)

7. Telephone hackers were originally called _____.
   a. dead heads
   b. cereal eaters
   c. phone phreaks
   d. script kiddies

8. The out-of-band signaling system was designed primarily for redundancy, meaning that the out-of-band range can be used to carry voice if needed. True/False

9. The out-of-band signaling system is built around the _____ network.

 a. ST5

 b. SS5

 c. ST7

 d. SS7

10. Out-of-band signaling allows signaling to _____ once a call has been set up.

 a. continue

 b. not continue

 c. be hacked over the voice network

 d. transmit files/data from customer computer to computer

11. _____ are the switches at both of the caller and receiver ends in an out-of-band signaling system.

 a. Signal control points (SCPs)

 b. Signal switching points (SSPs)

 c. Signal transfer points (STPs)

 d. SS7s

12. _____ are the out-of-band enabled switches that handle the long-distance path of the out-of-band signal traffic.

 a. Signal control points (SCPs)

 b. Signal switching points (SSPs)

 c. Signal transfer points (STPs)

 d. SS7s

13. _____ are the databases that provide information on advanced call-processing capabilities in an out-of-band signaling network.

 a. Signal control points (SCPs)

 b. Signal switching points (SSPs)

 c. Signal transfer points (STPs)

 d. SS7s

14. Redundantly installed STPs in an out-of-band network configuration are commonly referred to as _____.

 a. tied

 b. connected

 c. mated

 d. networked

15. The universal schematic symbol for an out-of-band signaling SSP is an _____.

 a. open square

 b. open circle

 c. open circle with a diagonal line through it (top right to bottom left)

 d. open square with a diagonal line through it (top right to bottom left)

16. The universal schematic symbol for an out-of-band signaling STP is a _____.

 a. Open square

 b. Open circle

 c. Open circle with a diagonal line through it (top right to bottom left)

 d. Open square with a diagonal line through it (top right to bottom left)

17. Interconnected lines between redundant out-of-band signaling devices are referred to as _____ links.

 a. voice

 b. PC to PC

 c. signal

 d. telephone

18. Interconnected lines between redundant out-of-band signaling devices typically communicate at _____ or _____. Please select the two correct answers.

 a. 28.8 Kbps

 b. 56 Kbps

 c. 64 Kbps

 d. 128 Kbps

Section 6.2

19. In the United States only the local loop is typically still in analog format. True/False

20. _____ is considered all digital including the local loop.

    a. POTS

    b. RBOC

    c. ISDN

    d. CLEC

21. The _____ is the organization that sets the world standard for the ISDN.

    a. FCC

    b. NSA

    c. ITU

    d. Lucent

22. The ISDN _____ channels are used to transmit digital (nonsignal) information.

    a. A-

    b. B-

    c. C-

    d. D-

23. The ISDN _____ channels are used to transmit signaling information.

    a. A-

    b. B-

    c. C-

    d. D-

24. _____ is required for ISDN to work with the ISDN D-channel integrated directly into the network.

    a. ST5

    b. SS5

    c. ST7

    d. SS7

25. The ISDN uses out-of-band signaling on the local loop. True/False

26. POTS uses out-of-band signaling on the local loop. True/False

27. Call setup over the ISDN is relatively long when compared to call setup over a POTS line. True/False

28. The ISDN basic-rate interface (BRI) is also commonly referred to as _____ service.

    a. 23B+D

    b. B+D

    c. 2B+D

    d. B+2D

29. The ISDN basic-rate interface (BRI) uses _____ data-transmission channels.

    a. 16-Kbps

    b. 56-Kbps

    c. 64-Kbps

    d. 256-Kbps

30. In the United States, telephone companies provide a _____ interface wire pair for connecting the customer location to the ISDN switch.

    a. T

    b. U

    c. V

    d. W

31. At a customer ISDN location, a/an _____ is used to convert the two-wire incoming pair to a four-wire on-site configuration.

    a. S/T

    b. NT-1

    c. low-pass filter

    d. bridge tap

32. A single ISDN S/T interface can have up to _____ ISDN devices attached and sharing a single ISDN connection.

    a. two

    b. three

    c. five

    d. seven

33. The ISDN primary-rate interface (PRI) is also commonly referred to as _____ service.

    a. 23B+D

    b. B+D

    c. 2B+D

    d. B+2D

34. The ISDN primary-rate interface (PRI) uses _____ data-transmission channels.

    a. 16-Kbps

    b. 56-Kbps

    c. 64-Kbps

    d. 256-Kbps

35. The European ISDN primary-rate interface (PRI) provides _____ B-channels.

    a. two

    b. 23

    c. 30

    d. 64

36. ISDN allows the aggregation of B-channels into _____ channels.

    a. D-

    b. A-

    c. J-

    d. H-

Section 6.3

37. The ISDN BRIs and PRIs are sometimes referred to as _____ because of the bandwidth limitations of the copper wire local loop.

    a. narrowband

    b. broadband

    c. notchband

    d. low pass

38. Broadband ISDN uses the high-bandwidth capabilities of _____ to overcome copper wire bandwidth limitations.

    a. load coils

    b. coaxial cable

    c. infrared wireless transmission

    d. fiber optics

Discussion Questions

Section 6.1

1. Describe the process a telephone system hacker would use to obtain free long-distance service using the old in-band signaling telephone network.

2. Why is the SS7 network faster than the old in-band signaling method?

3. What is meant by redundancy in the current SS7 long-distance network?

4. Describe the function and interconnection of SSPs, STPs, and SCPs in the SS7 network.

Section 6.2

5. Describe the difference between a POTS residential connection and an ISDN residential connection.

6. What organization controls worldwide telecommunications standards?

7. For what are the ISDN B-channels used? What are the ISDN D-channels used?

8. Does ISDN use in-band or out-of-band signaling on the local loop? What are the advantages?

9. Describe ISDN BRI. How much bandwidth is provided? Can it all be used for data? Why or why not?

10. What is the ISDN interface/connection provided by U.S. telephone companies called?

11. For what is an ISDN NT-1 used?

12. Describe ISDN PRI in the United States. How much bandwidth is provided? Can it all be used for data? Why or why not?

13. What is the difference between European and U.S. ISDN PRI?

Section 6.3

14. What is the major difference between ISDN BRI or PRI and broadband ISDN? Do you think it will ever be available in people's homes?

15. Research broadband ISDN on the Web. Is it widely used in the United States? How about other countries?

Data on the Legacy Network

Objectives   Upon completion of this chapter, the student should be able to:

- Understand how the existing PSTN is used to transmit and receive data.
- Describe how an analog modem works.
- Discuss the three primary functions of a modem.
- Comprehend and describe the differences between four different types of modulation (ASK, FSK, PSK, and QAM).
- Define how noise is calculated and reported in a transmission system.
- Determine the maximum data bit rate given a transmission system bandwidth and signal-to-noise ratio.
- Describe different analog modem modulation standards.
- Describe what a defacto standard is.
- Understand odd and even parity checking.
- Define different methods of modem error detection and correction.
- Understand how and why modems use data compression.

Outline   7.1   Analog Modems

7.2   Noise

7.3   Modulation Standards

7.4   The Digital Connection

7.5   Error Correction

7.6   Data Compression

Key Terms

acknowledgment (ACK)

American Standard Code for Information Interchange (ASCII)

amplitude-shift keying (ASK)

automatic retransmission request (ARQ)

baud

cyclic redundancy check (CRC)

data compression

decibel (dB)

defacto standard

demodulation

frequency-shift keying (FSK)

Huffman encoding

Internet service provider (ISP)

modulation

no acknowledgment (NAK)

noise

parity checking

phase-shift keying (PSK)

quadrature amplitude modulation (QAM)

Shannon's law

signal-to-noise ratio (SNR)

Introduction

We now have a good understanding of the existing public switched telephone network (PSTN) and can look at how we modulate and demodulate digital data and transmit it across the existing voice designed network.

7.1 Analog Modems

Analog modems are used in a home or business allowing the local loop, with its limited audible bandwidth of 300 to 3300 hertz, to be used to transmit and receive data. An analog modem is a device that takes a digital signal from a computer or other digital device and converts or modulates it into an analog voice frequency for transmission on the PSTN. At the receiving end, the analog signal is then demodulated or converted back to digital by the receiving modem for the receiving digital device. Data-transmission rates are typically expressed in both **baud** and bits per second (bps). One baud is the data communications equivalent of one hertz in that it represents one electronic state change per second. The term *baud* is named after the French engineer Jean-Maurice-Emile Baudot (1845–1903), who in 1874 received a patent on a telegraph code that replaced Morse code in the mid-20th century.

Modem Function

Today, using modern modulation techniques, one state change usually includes more than one bit of data and the bit-per-second (bps) unit is commonly used. Analog modems have three primary functions:

1. Modulation and demodulation (e.g., V.90)
2. Error correction (e.g., MNP4 and V.42)
3. Data compression (e.g., V.42bis and MNP5)

Modulation

Modulation is defined as the conversion of digital signals to analog signals on a sending communications device, and **demodulation** is defined as the conversion of analog signals to digital signals that a receiving device can understand. Signals can be modulated in many different ways; but, in all techniques, binary signals (digital 0s and 1s) are converted to analog *audiofrequency (AF)* tones. On the receiving end, these AF tones are demodulated or converted back to binary signals for the receiving digital device. The modulation and demodulation process is illustrated in figure 7-1. To do this, a modem generates an *analog carrier* or *carrier signal*. This signal is then altered or modulated in different ways to carry the information to the receiving device. These carrier signals are modulated in one of four ways:

1. Amplitude-shift keying (ASK)
2. Frequency-shift keying (FSK)
3. Phase-shift keying (PSK)
4. Quadrature amplitude modulation(QAM)

FIGURE 7-1 Model modulation and demodulation.

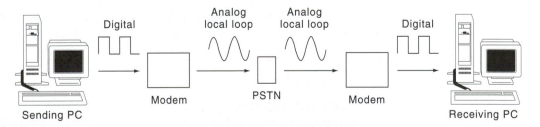

FIGURE 7-2
Amplitude-shift
keying (ASK).

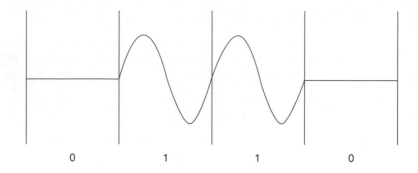

FIGURE 7-3
Frequency-shift
keying (FSK).

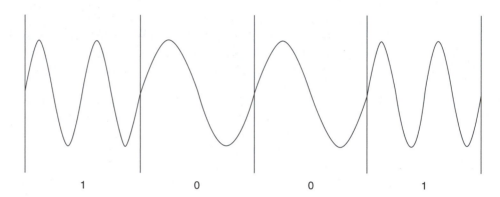

Amplitude-Shift Keying

Amplitude-shift keying (ASK) is a modulation technique that varies only the *amplitude* of the analog carrier signal. The frequency and phase remain constant. Basically, the signal is turned on and off, as illustrated in figure 7-2. ASK is also referred to as *OOK (on-off-keying).*

Frequency-Shift Keying

Frequency-shift keying (FSK) is a modulation technique that varies only the *frequency* of the analog carrier signal. As illustrated in figure 7-3, the amplitude and phase stay constant.

Phase-Shift Keying

Phase-shift keying (PSK) is a modulation technique that shifts only the *phase* of the carrier signal. A common variation is *binary-phase-shift keying (BPSK),* in which only two different phases are used. BPSK is illustrated in figure 7-4.

Each modulation technique described so far works well individually; but, with the limited 3000-hertz bandwidth of the local loop, each does not individually provide enough throughput for today's Internet-based data world. How do we get more throughput? Multiple bits per cycle are transmitted by

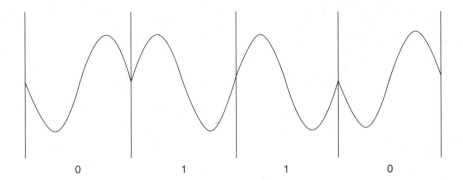

FIGURE 7-4 Binary phase-shift keying (BPSK).

combining different modulation methods. The most effective way to do this is to use **quadrature amplitude modulation (QAM).**

Quadrature Amplitude Modulation

Quadrature amplitude modulation (QAM) is a modulation technique that combines amplitude-shift keying and phase-shift keying to increase data throughput in a transmission system. The relationship between phase and amplitude is typically represented with a *constellation diagram*. We can look at QAM 16 as an example. QAM 16 provides four bits, or quadbits, per cycle, so:

Data-transmission rate = 4 × baud rate

The *16* comes from the fact that there are four bits transmitted per baud and $2^4 = 16$. The constellation diagram for QAM 16 is shown in figure 7-5. Each symbol on the diagram represents a specific amplitude and frequency. With QAM 16 three amplitudes are used—.31 volt, .85 volt, and 1.16 volts—and 12 phase shifts—15 degrees, 45 degrees, 75 degrees, 105 degrees, 135 degrees,

FIGURE 7-5 QAM 16.

```
1101 ●     ● 1100 │ 1110 ●     ● 1111

1001 ●     ● 1000 │ 1010 ●     ● 1011
───────────────────┼───────────────────
0001 ●     ● 0000 │ 0010 ●     ● 0011

0101 ●     ● 0100 │ 0110 ●     ● 0111
```

FIGURE 7-6
QAM 16,
1110 phasor
representation.

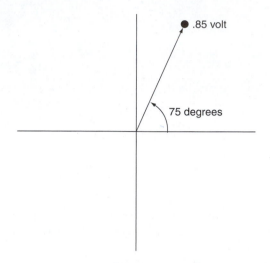

FIGURE 7-7
QAM 16,
1000 phasor
representation.

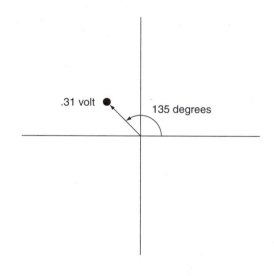

165 degrees, 195 degrees, 225 degrees, 255 degrees, 285 degrees, 315 degrees, and 345 degrees. A phasor representation of the constellation *spectral point* 1110 is shown in figure 7-6. The spectral point in the figure represents an amplitude of .85 volt at an angle of 75 degrees.

A phasor representation of the 1000 constellation spectral point is shown in figure 7-7. This point represents an amplitude of .31 volt at an angle of 135 degrees.

If we look at constellation diagrams closely, we can see that symbols can potentially overlap and there is a potential to get "fuzzy" if amplitudes and phases are not precise. For example, in figure 7-8, we can look at a zoomed-

FIGURE 7-8
Intersymbol
interference.

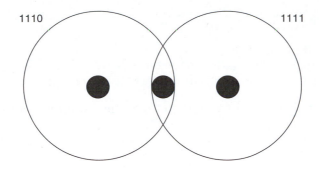

in view of the symbol points for the *1110* and *1111*. The middle point here represents a symbol point that can be interpreted as either a bit pattern of *1110* or *1111* because the symbol point appears between the two reference points. This type of symbol overlapping is called *intersymbol interference*. Recall that Henry Nyquist's sampling theorem states that an analog signal can be uniquely reconstructed, without *error,* from samples taken at equal *time* intervals if the *sampling rate* is equal to or greater than twice the highest *frequency component* in the analog signal. Expressed mathematically, Nyquist's sampling theorem says:

Sampling rate = 2 × highest frequency

We can now work this equation a little deeper and say that it also represents the maximum data rate (C in bits per second) in a noiseless channel that can be supported using a given highest frequency.

C = 2 × highest frequency

This equation assumes that we are transmitting at one bit per baud or cycle. We know that the public switched telephone network has a bandwidth, if we include the guard bands, of approximately 4000 hertz. This 4000 hertz also represent the highest transmitted frequency. Using this value and Nyquist's sampling theorem, we can calculate the maximum channel capacity of the public switched telephone network as follows:

C = 2 × bandwidth = 2 × 4000 Hz = 8000 bps

Remember, using modulation techniques like QAM, modems can now transmit multiple bits per baud or cycle. Taking this into consideration, Nyquist's sampling theorem is modified and becomes:

$C = 2(\text{bandwidth})\log_2 M$

Where M is the number of combinations that can be detected. Consider example 7.1.

●—**EXAMPLE 7.1**

●—**Problem**

A common voice grade circuit has a bandwidth of 3.4 kilohertz and can use QAM 16. Find the maximum data rate using Nyquist's equation.

●—**Solution**   Before we start to solve this example, recall $\log_2$ by looking at an example.

Recall:

$\log_2 3 = 2$

This means:

$2^x = 3$

How do we calculate x if our calculators do not directly calculate base 2 logs? We use the following:

$$x = \frac{\log_{10} 3}{\log_{10} 2} = 1.5849$$

To check, does $2^x = 3$?

$2^{1.5849} = 3$

Yes, so the answer is correct. Now, let us return to our maximum data rate example:

Bandwidth = 3400 Hz $M = 16$

$C = 2(\text{bandwidth}) \log_2 M = 2(3400 \text{ Hz}) \log_2 16 = 2(3400 \text{ Hz})(4)$
   $= 27{,}200$ bps

In this example, the maximum bit rate is 27,200 bits per second.

7.2 Noise

Nyquist's equation is a theoretical maximum because it does not take **noise** into account. Noise is any part of a signal that is not the actual desired signal for transmission. Noise is typically measured and indicated in the form of the **signal-to-noise ratio (SNR).** The signal-to-noise ratio (SNR) is a ratio of the actual signal to the amount of noise included along with the signal and is expressed in **decibels (dB).** The signal-to-noise ratio (SNR) of a transmission system is defined by the following equation:

$$SNR \text{ in dB} = 10 \log_{10}\left(\frac{\text{signal power}}{\text{noise power}}\right)$$

or

$$SNR \text{ in dB} = 10 \log_{10}\left(\frac{S}{N}\right)$$

where S = signal power and N = noise power.

Note that a high SNR is favorable to a lower one. The higher the SNR in decibels is, or $S{:}N$ is as a ratio, the better the transmission system quality will be. These values indicate a ratio of relative signal power to noise power, and the greater the signal is when compared to noise, the better the quality of the transmitted signal is at the receiving end. Consider example 7.2.

●—EXAMPLE 7.2

●—Problem

Signal voltage of a communications system is measured at 45 volts$_{rms}$, and signal noise on the same system is measured at .9 volt$_{rms}$. Calculate the SNR in decibels.

●—Solution

S = 45 volts$_{rms}$

N = .9 volt$_{rms}$

$$SNR \text{ in dB} = 10 \log_{10}\left(\frac{S}{N}\right) = 10 \log_{10}\left(\frac{45 \text{ volts}_{rms}}{.9 \text{ volt}_{rms}}\right) = 17 \text{ dB}$$

In example 7.2, we calculated the SNR given the signal and noise levels. If given the SNR of a system, we can calculate the $S{:}N$ value, as shown in example 7.3.

●—EXAMPLE 7.3

●—Problem

A system has an *SNR* of 33 decibels. Calculate the *S:N* value.

●—Solution

$$SNR \text{ in dB} = 10 \log_{10}\left(\frac{S}{N}\right)$$

$$33 \text{ dB} = 10 \log_{10}\left(\frac{S}{N}\right)$$

Solving for *S:N*,

$$\frac{33 \text{ dB}}{10} = \log_{10}\left(\frac{S}{N}\right)$$

$$3.3 \text{ dB} = \log_{10}\left(\frac{S}{N}\right)$$

$$10^{3.3} = \frac{S}{N}$$

$$1995.26 = \frac{S}{N}$$

Shannon's Law

Shannon's law is a fundamental law that states that the maximum data bit rate is determined by the bandwidth and the SNR of a communications system. In mathematical form:

$$C = (\text{bandwidth}) \log_{2}\left(1 + \frac{S}{N}\right)$$

Notice we are using the signal to noise ratio *S:N* as a straight ratio here; we are not using SNR in decibels. This equation defines the local loop very well. We already have a system designed for relatively low bandwidth; and, on top of that, we have noise to worry about. In high-noise situations, the noise problem is solved by reducing bandwidth. This results in less noise being passed to the receiving circuitry with a reduced bit rate trade-off.

7.3 Modulation Standards

Over the years, different modulation standards have become popular with each new generation producing higher bandwidths for the communicating devices. Some so-called standards were not standards at all—they were *proprietary;* and, for two devices to talk to each other, the devices either had to be from the same manufacturer or had to be from different companies with one of the companies having permission from the other to use the **proprietary standard.** Others are referred to as **defacto standards.** A defacto standard is a standard that has not been approved by a standards organization but has been accepted and used by the industry as a standard. The modem company Hayes set the most widely used defacto standard with a set of commands for controlling modems. Other popular defacto standard examples include the Xmodem communications protocol and Hewlett-Packard's PostScript laser printer control language. Following is a brief list of the different modulation standards:

Bell 103

An old full-duplex standard that communicates 1 bit per baud or cycle at 300 bits per second. Uses FSK to modulate signals.

V.21

Used in Europe and Japan and is considered an international standard. Communicates full duplex at 300 bits per second and is similar to the Bell 103 standard but is not Bell 103 compatible.

Bell 212A

An older full-duplex standard that communicates at the lower Bell 103 standard rate and also communicates at the high-speed rate of 2 bits per baud or cycle. In high-speed mode, it is capable of 1200 bits per second at 600 baud. Uses FSK modulation when communicating with Bell 103 modems and PSK to communicate with other Bell 212A modems in high-speed mode.

V.22

This standard is similar to V.21 but can communicate at 1200 bits per second.

V.22bis

Uses PSK to communicate full duplex at the lower bit rate of 1200 bits per second and QAM to communicate full duplex at 4 bits per baud or cycle at 2400 bits per second. Used in the United States and in other countries. *Bis means twice* in Latin and, used here, is short for *second generation.*

V.32

A full-duplex standard operating at 9600 bits per second, with a 2400-baud rate. This standard incorporates both error correction and control and, as a result, works well on noisy phone lines. Uses QAM to modulate signals and Trellis encoding to split the transmitted bit stream into groups of five bits called *quintbits.* Each quintbit is represented as a constellation point on the QAM diagram.

V.32bis

Second-generation V.32. Transmits full-duplex 6 bits per baud or cycle, providing bit rates of 14,400 bits per second. Will automatically fall back to V.32 if the phone line becomes noisy and will fall forward when noise goes away.

V.34

Full-duplex standard that provides, without data compression, a bit rate of 28,800 bits per second.

V.34+

Many V.34 modems came with flash ROM BIOSs and, with a V.34+ upgrade they can communicate at two higher bit rates, 31.2 kilobits per second and 33.6 kilobits per second without data compression.

V.90

Combined the technologies of two proprietary standards, the U.S. Robotics X2 and Rockwell's K5 flex technology. Both U.S. Robotics and Rockwell were selling these proprietary standards before the V.90 standard with most modems having flash ROM BIOSs that could easily be upgraded.

V.90 modems are asymmetrical devices because they transmit at different bit rates in the upstream and downstream directions. Upstream traffic is limited to the V.34+ maximum bit rate of 33.6 bits per second, while downstream bit rates are limited to a bit rate of 53 kilobits per second by the FCC. The FCC will not allow the 53 kilobits-per-second rate to be exceeded because of the potential of crosstalk at the higher bit rates.

V.92

The V.92 standard is a souped-up version of the V.90 standard. V.92 adds a *fast connection* to the modem-to-modem handshake negotiation process most of us are used to hearing when a modem first connects. Manufacturers realize that most Internet modem calls are made to the same *Internet service provider (ISP)* number and use the same username and password (ISPs are discussed further in section 7.4 of this chapter). The V.92 modem stores these ISP connection settings within the modem and reuses them each time a dial-up connection is made. For single-phone-line users (voice and data on one analog line), the V.92 standard also provides better support for call waiting, allowing ISPs to set how long an Internet connection can be placed on hold before the connection is dropped. The V.90 standard uses a fixed hold time of approximately 7 seconds, which, in almost all cases, is not enough time to pick up an incoming call without the Internet connection line dropping.

7.4 The Digital Connection

Both V.90 and V.92 modems were designed for Internet use over the PSTN and depend on a portion of an ISP connection to be digital. If a portion of the connection is digital, an *analog-to-digital (A/D)* conversion is not required in the downstream direction from the ISP. This conversion elimination reduces overall noise on the line, because there is no *quantization effect*.

Let's look at both cases (with and without A/D conversion) a little more closely.

With Analog-to-Digital Conversion

An A/D conversion is required when the ISP is connected to the telephone network with a simple local loop connection, as illustrated in figure 7-9. The figure assumes that the ISP is not connected digitally (T-1, etc.) to the telephone network, and an analog-to-digital (A/D) conversion is required to move the data from the ISP into the telephone network. This A/D conversion requires quantization, and this quantization produces a relatively high level of noise. We can look at how this noise affects the system by calculating the channel capacity using Shannon's law:

$$C = \left(\text{bandwidth}\right) \log_2\left(1 + \frac{S}{N}\right)$$

FIGURE 7-9 Analog connection at ISP.

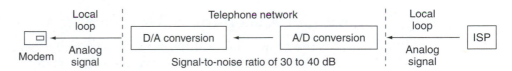

•—EXAMPLE 7.4

•—Problem

We know the usable bandwidth of the local loop is approximately 3000 hertz, and we are given a signal-to-noise ratio (SNR) of 30 to 40 decibels. Calculate the channel capacity using Shannon's law.

For our example, let's use an SNR of 40 decibels.

•—Solution
Before we can plug the numbers into Shannon's law, we must convert the SNR in decibels to the S:N ratio.

$$SNR \text{ in dB} = 10 \log_{10}\left(\frac{S}{N}\right)$$

$$40 \text{ dB} = 10 \log_{10}\left(\frac{S}{N}\right)$$

Solving for S:N,

$$\frac{40 \text{ dB}}{10} = \log_{10}\left(\frac{S}{N}\right)$$

$$4 \text{ dB} = \log_{10}\left(\frac{S}{N}\right)$$

$$10^4 = \frac{S}{N}$$

$$10,000 = \frac{S}{N}$$

Now we can substitute into Shannon's law as follows:

$$C = (\text{bandwidth}) \log_2\left(1 + \frac{S}{N}\right) = (3000 \text{ Hz}) \log_2 (1 + 10,000)$$

$$C = (3000 \text{ Hz})\left(\frac{\log_{10} 10,001}{\log_{10} 2}\right) = 3000 \text{ Hz}\left(\frac{4}{.3010}\right) = 39,864 \text{ bps}$$

Example 7.4 represents a calculated maximum for data traffic on the PSTN that has to be converted from analog to digital.

Without Analog-to-Digital Conversion

With the growth of the Internet since the mid-1990s, most modems are no longer being used for PC-to-PC-type communications. Consumers are dialing

FIGURE 7-10
Internet service provider (ISP) point of presence (POP).

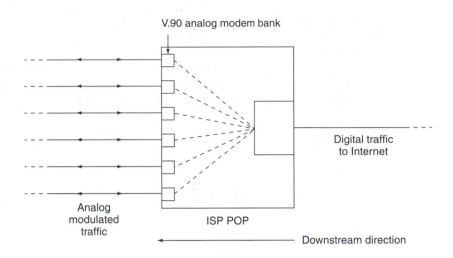

V.90 analog modem bank

Digital traffic to Internet

Analog modulated traffic

ISP POP

Downstream direction

FIGURE 7-11
Digital ISP connection.

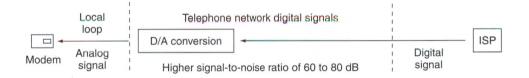

Local loop

Telephone network digital signals

Modem

D/A conversion

ISP

Analog signal

Digital signal

Higher signal-to-noise ratio of 60 to 80 dB

into companies called *Internet service providers (ISPs)* that provide customer Internet access and are using their modems to access the Internet. The ISPs commonly have point of presence (POP) set up in different service areas. The ISP POPs, as discussed in an earlier chapter, are a series of modems installed in a modem bank that customers dial into using the PSTN. The ISP then runs a digital line (T-1, etc.) from the POP that carries all customer Web traffic to and from the POP. This is illustrated in figure 7-10. Notice in the figure that an analog-to-digital (A/D) conversion is not required for traffic moving in the downstream direction. The digital-to-analog (D/A) conversion is the only one required to get the traffic in the proper format for the local loop. Without this downstream A/D conversion, the noise associated with quantization is eliminated and the SNR increases by a factor of 50 percent. Remember that a high signal-to-noise ratio indicates more signal when compared to the noise level, so the higher the better. This is further illustrated in figure 7-11.

Example 7.5 shows how this higher signal-to-noise ratio affects transmission. Compare these results to example 7.4.

—EXAMPLE 7.5

—Problem

We know the usable bandwidth of the local loop is approximately 3000 hertz, and we are given a higher signal-to-noise ratio (SNR) of 60 to 80 decibels. Calculate the channel capacity using Shannon's law.

—Solution

For our example, let us use an SNR of 70 decibels. Before we can plug the numbers into Shannon's law, we must convert the SNR in decibels, to the $S{:}N$ ratio:

$$SNR \text{ in dB} = 10 \log_{10}\left(\frac{S}{N}\right)$$

$$70 \text{ dB} = 10 \log_{10}\left(\frac{S}{N}\right)$$

Solving for $S{:}N$,

$$\frac{70 \text{ dB}}{10} = \log_{10}\left(\frac{S}{N}\right)$$

$$7 \text{ dB} = \log_{10}\left(\frac{S}{N}\right)$$

$$10^7 = \frac{S}{N}$$

$$10,000,000 = \frac{S}{N}$$

Now we can substitute into Shannon's law as follows:

$$C = (\text{bandwidth}) \log_2\left(1 + \frac{S}{N}\right) = (3000 \text{ Hz}) \log_2 (1 + 10,000,000)$$

$$C = (3000 \text{ Hz})\left(\frac{\log_{10} 10,000,001}{\log_{10} 2}\right) = 3000 \text{ Hz}\left(\frac{7}{.3010}\right) = 69,760 \text{ bps}$$

Example 7.5 represents a calculated maximum for data traffic on the PSTN that does not have to be converted from analog to digital in the downstream direction. Notice that it is much greater than the bit rate calculated in example 7.4. Modem manufacturers and ISPs have taken advantage of this configuration and now offer V.90 and V.92 modem support in most locations.

7.5 Error Correction

In noisy telephone line environments, error correction is critical to ensure data integrity. Small amounts of noise can be ignored in normal voice conversations but any amount of noise that could potentially turn a binary *1* into a *0* or a *0* into a *1* has the potential to destroy the integrity of a piece of data being sent. **Parity checking** is a technique used for data-transmission error checking and was the earliest method used to detect data-transmission errors.

Parity Checking

The **American Standard Code for Information Interchange (ASCII)** is a common format used to transmit data between computers with all characters represented using a seven-bit binary number. The ASCII format was devised by *American National Standards Institute (ANSI:* http://www.ansi.org), a private, nonprofit organization that administers and coordinates the U.S. voluntary standardization and conformity assessment system. The ASCII format is a way to represent each alphabet character, number and some nonalphanumeric characters using a *seven-bit* binary code. The ASCII representation is shown in table 7-1. Notice that seven bits are used to represent the 128 listed values. The eighth bit is used for parity checking. We next review two types of parity checking, odd and even.

Odd Parity Checking

Odd parity checking works by counting the number of *1*s in a seven-bit value. If the number of *1*s is even, then the eighth bit is made a *1* to force the number of *1*s to be an odd number count. Following are some odd parity examples.

EXAMPLE 7.6

●—Problem
Set the parity bit for the small letter *z* for odd parity.

●—Solution   Referencing the ASCII table:

$z = 1111010$

The number of *1*s in the seven-bit binary representation is five, which is an odd number. Therefore, the parity bit is set to *0* and the eighth bit representation of the letter *z* with odd parity is *01111010* with the first bit position on the left indicating parity. This is illustrated in figure 7-12.

TABLE 7-1 ASCII representation.

Decimal	Octal	Hex	Binary	Value
000	000	000	00000000	NUL (null char.)
001	001	001	00000001	SOH (start of header)
002	002	002	00000010	STX (start of text)
003	003	003	00000011	ETX (end of text)
004	004	004	00000100	EOT (end of transmission)
005	005	005	00000101	ENQ (enquiry)
006	006	006	00000110	ACK (acknowledgment)
007	007	007	00000111	BEL (bell)
008	010	008	00001000	BS (backspace)
009	011	009	00001001	HT (horizontal tab)
010	012	00A	00001010	LF (line feed)
011	013	00B	00001011	VT (vertical tab)
012	014	00C	00001100	FF (form feed)
013	015	00D	00001101	CR (carriage return)
014	016	00E	00001110	SI (serial in)(shift out)
015	017	00F	00001111	SO (serial out)(shift out)
016	020	010	00010000	DLE (data link escape)
017	021	011	00010001	DC1 (XON) (device control 1)
018	022	012	00010010	DC2 (device control 2)
019	023	013	00010011	DC3 (XOFF) (device control 3)
020	024	014	00010100	DC4 (device control 4)
021	025	015	00010101	NAK (negative acknowledgement)
022	026	016	00010110	SYN (synchronous idle)
023	027	017	00010111	ETB (end of trans. block)
024	030	018	00011000	CAN (cancel)
025	031	019	00011001	EM (end of medium)

Decimal	Octal	Hex	Binary	Value
026	032	01A	00011010	SUB (substitute)
027	033	01B	00011011	ESC (escape)
028	034	01C	00011100	FS (file separator)
029	035	01D	00011101	GS (group separator)
030	036	01E	00011110	RS (request to send)(record separator)
031	037	01F	00011111	US (unit separator)
032	040	020	00100000	SP (space)
033	041	021	00100001	!
034	042	022	00100010	"
035	043	023	00100011	#
036	044	024	00100100	$
037	045	025	00100101	%
038	046	026	00100110	&
039	047	027	00100111	'
040	050	028	00101000	(
041	051	029	00101001	)
042	052	02A	00101010	*
043	053	02B	00101011	+
044	054	02C	00101100	,
045	055	02D	00101101	-
046	056	02E	00101110	.
047	057	02F	00101111	/
048	060	030	00110000	0
049	061	031	00110001	1
050	062	032	00110010	2
051	063	033	00110011	3

(continued)

TABLE 7-1
(*Continued*)

Decimal	Octal	Hex	Binary	Value
052	064	034	00110100	4
053	065	035	00110101	5
054	066	036	00110110	6
055	067	037	00110111	7
056	070	038	00111000	8
057	071	039	00111001	9
058	072	03A	00111010	:
059	073	03B	00111011	;
060	074	03C	00111100	<
061	075	03D	00111101	=
062	076	03E	00111110	>
063	077	03F	00111111	?
064	100	040	01000000	@
065	101	041	01000001	A
066	102	042	01000010	B
067	103	043	01000011	C
068	104	044	01000100	D
069	105	045	01000101	E
070	106	046	01000110	F
071	107	047	01000111	G
072	110	048	01001000	H
073	111	049	01001001	I
074	112	04A	01001010	J
075	113	04B	01001011	K
076	114	04C	01001100	L
077	115	04D	01001101	M

Decimal	Octal	Hex	Binary	Value
078	116	04E	01001110	N
079	117	04F	01001111	O
080	120	050	01010000	P
081	121	051	01010001	Q
082	122	052	01010010	R
083	123	053	01010011	S
084	124	054	01010100	T
085	125	055	01010101	U
086	126	056	01010110	V
087	127	057	01011111	W
088	130	058	01011000	X
089	131	059	01011001	Y
090	132	05A	01011010	Z
091	133	05B	01011011	[
092	134	05C	01011100	\
093	135	05D	01011101	]
094	136	05E	01011110	^
095	137	05F	01011111	_
096	140	060	01100000	'
097	141	061	01100001	a
098	142	062	01100010	b
099	143	063	01100011	c
100	144	064	01100100	d
101	145	065	01100101	e
102	146	066	01100110	f
103	147	067	01100111	g

(continued)

TABLE 7-1
(*Continued*)

Decimal	Octal	Hex	Binary	Value
104	150	068	01101000	h
105	151	069	01101001	i
106	152	06A	01101010	j
107	153	06B	01101011	k
108	154	06C	01101100	l
109	155	06D	01101101	m
110	156	06E	01101110	n
111	157	06F	01101111	o
112	160	070	01110000	p
113	161	071	01110001	q
114	162	072	01110010	r
115	163	073	01110011	s
116	164	074	01110100	t
117	165	075	01110101	u
118	166	076	01110110	v
119	167	077	01110111	w
120	170	078	01111000	x
121	171	079	01111001	y
122	172	07A	01111010	z
123	173	07B	01111011	{
124	174	07C	01111100	|
125	175	07D	01111101	}
126	176	07E	01111110	~
127	177	07F	01111111	DEL

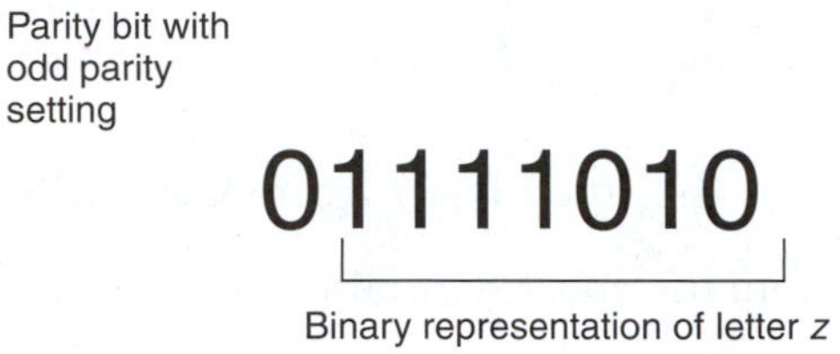

FIGURE 7-12 The letter *z* with odd parity.

Parity bit with odd parity setting

01111010

Binary representation of letter *z*

●—**EXAMPLE 7.7**

●—**Problem**

Set the parity bit for the small letter *q* for odd parity.

●—**Solution**   Referencing the ASCII table:

$q = 1110001$

The number of *1*s in the seven-bit binary representation is four, which is an even number. Therefore, the parity bit is set to *1* and the eighth bit representation of the letter *q* with odd parity is *11110001* with the last bit position on the left indicating parity. This is illustrated in figure 7-13.

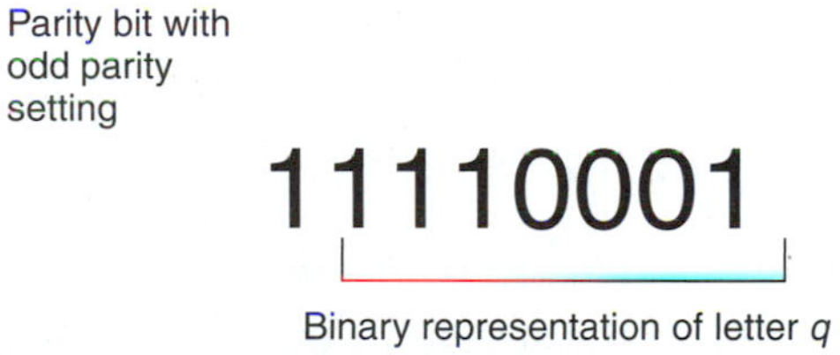

FIGURE 7-13 The letter *q* with odd parity.

Parity bit with odd parity setting

11110001

Binary representation of letter *q*

Even Parity Checking

Even parity checking works by counting the number of *1*s in a seven-bit value. If the number of *1*s is odd, then the eighth bit is made a *1* to force the number of *1*s to be an even number count. Following are some even parity examples.

●—**EXAMPLE 7.8**

●—**Problem**
Set the parity bit for the small letter z for even parity.

●—**Solution**   Referencing the ASCII table:

z = 1111010

The number of *1*s in the seven-bit binary representation is five, which is an odd number. Therefore, the parity bit is set to *1* and the eighth bit representation of the letter z with even parity is *11111010* with the last bit position on the left indicating parity, as illustrated in figure 7-14.

FIGURE 7-14 The letter z with even parity.

Parity bit with even parity setting

11111010

Binary representation of letter z

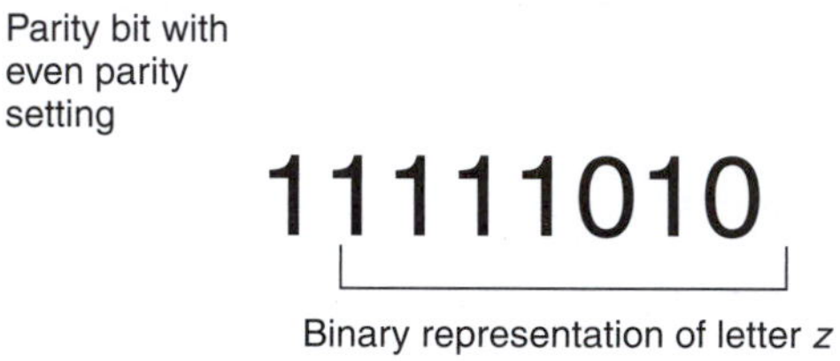

●—**EXAMPLE 7.9**

●—**Problem**
Set the parity bit for the small letter q for even parity.

●—**Solution**   Referencing the ASCII table:

q = 1110001

The number of *1*s in the seven-bit binary representation is four, which is an even number. Therefore, the parity bit is set to *0* and the eighth bit representation of the letter q with even parity is *01110001* with the last bit position on the left indicating parity. This is illustrated in figure 7-15.

FIGURE 7-15 The letter q with even parity.

Parity bit with even parity setting

01110001

Binary representation of letter q

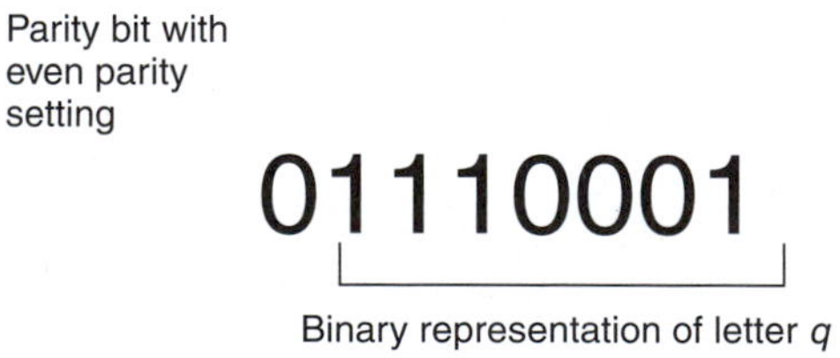

Parity checking is the error-checking method used for asynchronous serial, PC-to-PC communications over *RS-232 ports*. The devices are connected directly together; for communications to work properly, the ports on both devices must be set with the same type of parity checking. The actual error checking is handled by the communicating devices, and modems are not required to check parity. When using modems, parity checking is not used for error detection and correction; the modems themselves handle the detection and correction using modem error detection and correction protocols.

Modem Error Detection and Correction

Modem error-correction modes work by checking received data for errors at the receiving modem. If no errors are detected, data flows from the transmitting device to the receiving device at maximum bandwidth for the type of connection. If an error is detected by the receiving modem, the receiving modem makes an **automatic retransmission request (ARQ),** also referred to as an automatic repeat request, to the transmitting modem. Any retransmission requires bandwidth that could be used to send new data and, as a result, will reduce throughput and slow down overall transmission speeds. Two modems communicating with each other must use the same error-correction protocol for communications to work. The method modems use to error detect is called **cyclic redundancy checking (CRC).**

Cyclic Redundancy Checking

Cyclic redundancy checking (CRC) is a common method that modems use to detect transmission errors. CRC error detection involves an initial calculation on the sending modem. The two versions of the CRC protocol are, *CRC 16* and *CRC 32*.

CRC 16

Data is transmitted in blocks of 128 characters with each character the equivalent of a byte or eight individual bits. This means that a 128-character block is represented in binary form as a stream of 8×128 or 1024 bits. Treating this 1024-bit block of bits as a single number value can result in a huge number. We know from binary arithmetic that 1024 bits can represent the decimal numbers zero through ($2^{1024} - 1$). CRC 16 takes this large number and divides it by another large 17-bit prime number. Dividing by a prime number will always result in a remainder because prime numbers are only divisible by 1 and themselves without a remainder. Dividing a 1024-bit number by a 17 bit prime number will always result in a 16-bit remainder.

$$\frac{\text{1024-bit binary number}}{\text{17-bit prime number}} = \text{16-bit remainder}$$

This remainder is then added to the data packet as overhead and transmitted to the receiving modem. Upon receipt, the receiving modem pulls out the 1024 bits representing the 128 characters of data and performs the same calculation using the same 17-bit prime number. If the remainders match, the receiving modem assumes that the data has been delivered without errors. If the remainders do not match, the receiving modem will issue an *automatic retransmission request (ARQ)* to the sending modem. This automatic retransmission request (ARQ) is a reply back from a receiving terminal to the sender indicating that a piece of information that was sent needs to be retransmitted.

CRC 32

CRC 32 functions in the same way except for the size of the divisor and the resulting remainder. CRC 32 uses a 33-bit divisor resulting in a 32-bit remainder.

Automatic Retransmission Request

After performing the CRC error-detection calculation on each block of data, the receiving modem will transmit either an **acknowledgment (ACK)** or a **no acknowledgment (NAK).** An **acknowledgment (ACK)** is a reply back from a receiving terminal to the sender, indicating that a piece of information was successfully received. A **no acknowledgement (NAK)** is a reply back from a receiving terminal to the sender indicating that a piece of information was not successfully received. How these ACKs and NAKs are handled depends on the type of ARQ being used.

Discrete ARQ

Discrete ARQ works as indicated in figure 7-16. In the figure, the sending modem will send a block and wait for an ACK or a NAK to come back for the sent block. If an ACK is sent by the receiving modem to the sending modem, the sending modem will move on to the next block and send it. If a NAK is sent by the receiving modem to the sending modem, the sending modem will resend the previous block. Discrete ARQ is not an efficient way to transmit information when most of the information is being delivered without errors, because the sending modem must wait for a return ACK for each sent block before sending the next block.

Continuous ARQ

Continuous ARQ is a much more efficient way to transmit data on the existing telephone network. Continuous ARQ does not wait for an ACK to come back for each individual block. Blocks are sent with identification numbers, and the sending modem will continuously send blocks while receiving ACKs from the receiving modem. If the sending modem does receive a NAK or does not receive an ACK or a NAK for a specific block, the block is identified and

FIGURE 7-16
Discrete ARQ.

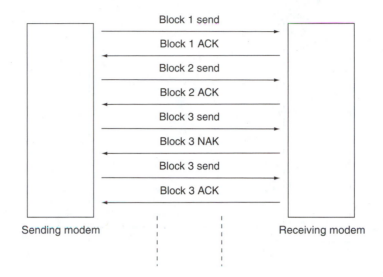

retransmitted. Some retransmission protocols will resend all blocks forward from a NAK, while others will use a selective ARQ protocol and only resend the selected bad block.

V.42

The current ITU standard for analog modem error detection and correction is V.42. A V.42 modem has the ability to use one of two individual protocols—the *microcom networking protocol (MNP)* or the *link access procedure for Modems (LAP-M)*. Protocol selection occurs on the *handshake* or initial communications configuration connection at the start of the modem-to-modem telephone call.

Microcom Networking Protocol

The MNP protocol uses a continuous ARQ method that will resend all blocks forward from a NAK. This process is outlined in figure 7-17. In the figure, blocks 1 through 4 are sent continuously. After block 4 has been sent, the sending modem receives a NAK on block 3 from the receiving modem. The sending modem backs up and sends block 3 and then block 4 (from the NAK block forward) again.

Link Access Procedure for Modems

The LAP-M protocol is a selective ARQ protocol and will only send individual NAK blocks back to the receiving modem. The process is outlined in figure 7-18. In the figure, blocks 1 through 4 are sent continuously. After block 4 has been sent, the sending modem receives a NAK on block 3 from the receiving modem. The sending modem backs up and resends only the bad block 3 and then moves on and sends block 5.

FIGURE 7-17 MNP
retransmission.

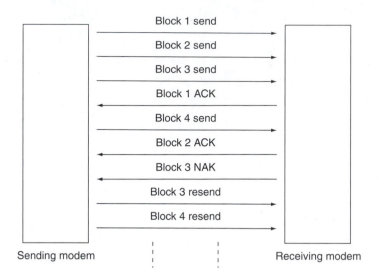

FIGURE 7-18 LAP-M
retransmission.

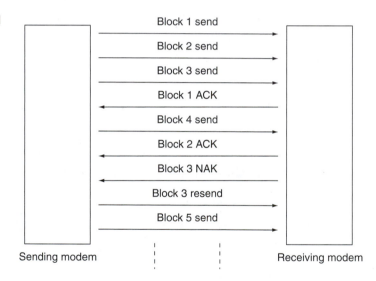

7.6 Data Compression

Data compression is used by modems to increase the amount of data transmitted (throughput) over set baud rates. Using data compression, the sending modem compresses the data into a compact form. The receiving modem uncompresses the data back to its original form, as illustrated in figure 7-19.

The effective data throughput is the communications rate between the sending and receiving devices. Data throughput depends directly on the bit

FIGURE 7-19
Modem data compression and uncompression.

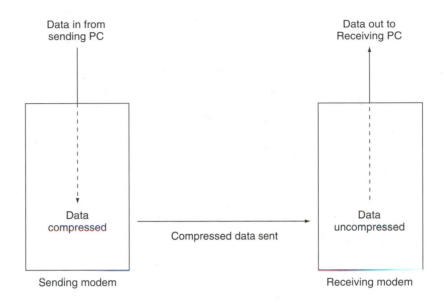

rate, on the amount of error retransmission, and on data compression. There are numerous data-compression algorithms, with **Huffman encoding** being the most fundamental. Huffman encoding is a data-compression algorithm that uses a *frequency table* and categorizes each *symbol,* or character, in a character string. The table may be constructed using the actual character string itself or from statistical data that represent the type of information that is to be compressed. For example, the frequency of different letters in the written English language has been studied and it has been found that the most common character is the *space* while the least common character is the letter *z.* A variable-length bit string is assigned to each character with the most common characters represented using fewer bits. Standard Huffman text encoding assigns three bits to represent the most common *space* character and 16 bits to represent the least common letter *z.*

We know that the *space* character in uncompressed format uses eight bits and in Huffman compressed format uses only three bits. We can see a realized compression ratio of 8:3. On the opposite end, we can know the *z* character in uncompressed format uses eight bits and in Huffman compressed format uses 16 bits. Here we have an inefficient compression ratio of 8:16. Huffman encoding is not concerned about this for the letter *z,* because the average text document does not use lots of *z*'s.

V.42bis Data-Compression Standard

In 1989, the ITU approved the V.42bis data-compression algorithm. The V.42bis standard requires V.42 error correction to also be used. The compression standard lists a maximum compression ratio of up to 8:1 but is typically

FIGURE 7-20
V.42bis dictionary
transmission.

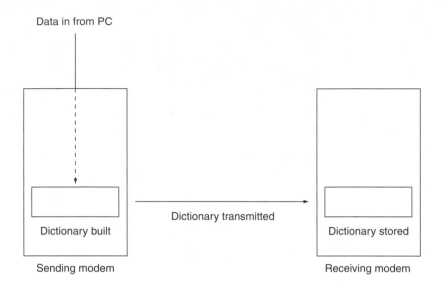

listed as providing a maximum compression ratio of 4:1 with a realistic ratio of 2.5:1. The algorithm builds a pattern dictionary on the sending modem, and this dictionary is sent to the receiving modem. The dictionary size ranges from 512 to 2048 bytes and is built by taking six characters at a time (more or fewer characters can be used) and mapping patterns for these six character strings. Figure 7-20 illustrates how the dictionary is transmitted.

The uncompressed character patterns are then compressed on the sending modem resulting in fewer bits being transmitted to represent the longer character strings, as illustrated in figure 7-21.

FIGURE 7-21
V.42bis data
transmission.

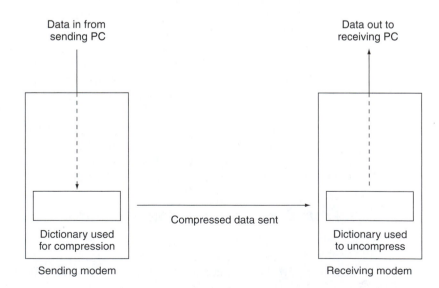

Data compression can occur only when the two communicating modems support the same compression method. If a modem supports more than one type of compression, the two modems will agree on the best compression technique that is common to both modems during the initial modem-to-modem communication or handshake process. Compression ratios can give false impressions of actual bit throughput rates. V.42bis compression is advertised as offering a compression ratio of up to 4:1. This compression ratio, when combined with a 28.8-kilobits-per-second bit rate, will produce throughput of 115.2 kilobits per second:

$$4 \times 28.8 \text{ Kbps} = 115.2 \text{ Kbps}$$

This is considered a maximum value by most modem manufacturers; a more realistic throughput value for a V.42bis modem communicating at a rate of 27.8 kilobits per second involves a compression ratio of 2.5:1. This translates to a throughput bit rate of 72 kilobits per second:

$$2.5 \times 28.8 \text{ Kbps} = 72 \text{ Kbps}$$

V.44 Data-Compression Standard

The latest data-compression standard approved by the ITU is V.44. This compression standard has been fine tuned to Internet traffic and provides a data-compression ratio in the range of 6:1. The earlier V.42bis data-compression standard provided data compression in the range of 4:1. The V.44 standard, when combined with the V.92 modem standard, produces data throughput rates in excess of 300 kilobits per second, which is fast when compared with the V.90/V.42bis throughput rates of 150 to 200 kilobits per second. Many ISPs made V.92/V.44 ports available to customers in mid-2001.

Summary

1. *Analog modems* have the voice network bandwidth of 300 to 3300 hertz as a limitation.

2. *One baud* represents one electronic state change per second.

3. Analog modems have four basic functions: *modulation, demodulation, error correction,* and *data compression.*

4. Analog modems *modulate,* or convert, computer digital signals to analog signals on the sending device.

5. Analog modems *demodulate,* or convert, analog signals to digital signals the receiving device can understand.

6. *Amplitude-shift keying (ASK)* is a modulation technique that varies the amplitude of a carrier signal.

7. *Frequency-shift keying (FSK)* is a modulation technique that varies the frequency of a carrier signal.

8. *Phase-shift keying (PSK)* is a modulation technique that varies the phase of a carrier signal.

9. *Quadrature amplitude modulation (QAM)* is a modulation technique that combines ASK and PSK on a carrier signal.

10. The *maximum channel capacity* of the PSTN is 8000 bits per second.

11. Noise is typically measured in the form of the *signal-to-noise ratio (SNR)* and expressed in *decibels (dB)*.

12. A *high SNR* is favorable to a low one.

13. Shannon's law states that the maximum data bit rate of a communications system is determined by the bandwidth and the SNR.

14. A *defacto standard* is a standard that has not been approved by a standards organization but has been accepted and used by the industry as a standard.

15. *V.90* and *V.92 modems* are designed for Internet use over the PSTN and depend on a portion of an Internet service provider's connection to be digital.

16. *Odd parity* and *even parity checking* are common ways to check digital transmissions between devices over serial communications ports.

17. Modems use an error-detection method called *cyclic redundancy check*ing (CRC).

18. *Automatic retransmission request (ARQ)* methods are used by analog modems to handle retransmissions when there are transmission errors between modem-connected devices.

19. Modems use *acknowledgments (ACK)* and *no acknowledgments (NAK)* to automate error detection and retransmission.

20. Analog modems use *compression* to increase transmission throughput.

Review Questions

Section 7.1

1. Analog modems are limited to an audible bandwidth of _____.
   - a. 4000 Hz to 1.1 MHz
   - b. 3000 Hz to 4000 Hz
   - c. 1000 Hz to 3300 Hz
   - d. 300 Hz to 3300 Hz

2. One baud is the data communications equivalent of _____.
   - a. 1 GHz
   - b. 1 MHz
   - c. 1 Hz
   - d. 1 KHz

3. The process of _____ converts audio-frequency signals to digital signals.
   - a. signaling
   - b. modulation
   - c. demodulation
   - d. switching

4. _____ converts digital signals to analog audiofrequency signals.
   - a. Signaling
   - b. Modulation
   - c. Demodulation
   - d. Switching

5. _____ varies only the amplitude of an analog carrier signal.
   - a. FSK
   - b. QAM
   - c. ASK
   - d. PSK

6. _____ varies only the frequency of an analog carrier signal.
   - a. FSK
   - b. QAM
   - c. ASK
   - d. PSK

7. _____ varies only the phase of an analog carrier signal.

a. FSK

b. QAM

c. ASK

d. PSK

8. _____ varies both the phase and the amplitude of an analog carrier signal.

a. FSK

b. QAM

c. ASK

d. PSK

9. A _____ diagram is used to indicate the relationship between phase and amplitude in a communications system.

a. star

b. sun

c. constellation

d. phase

Section 7.2

10. Nyquist's sampling theorem states that an analog signal can be uniquely reconstructed from a digitized signal from samples taken at a rate equal to or greater than _____ the highest frequency component in the analog signal.

a. twice

b. three times

c. half

d. one-third

11. Using Nyquist's sampling theorem, we can calculate a maximum channel capacity of the PSTN of approximately _____ bps.

a. 300

b. 3300

c. 4000

d. 8000

12. The SNR is typically expressed in _____.

a. amperes

b. decibels

c. volts

d. watts

13. Shannon's law states that the maximum bit rate of a communications system is determined by the _____.

a. voltage and the SNR

b. bandwidth and the SNR

c. bandwidth and the voltage

d. SNR and the voltage

Section 7.3

14. A _____ standard is a standard that has not been approved by any standards organization but has been accepted and used by the industry as a standard.

a. new

b. legacy

c. defacto

d. foreign

Section 7.4

15. These analog modem standards were designed for Internet use over the PSTN. Choose all correct answers.

a. V.22

b. V.22bis

c. V.90

d. V.92

Section 7.5

16. The earliest method used to detect data-transmission errors is called _____.

a. CRC 16

b. parity checking

c. CRC 32

d. ARQ

17. Both CRC-16 and CRC-32 error detection transmit data in blocks of _____ characters.

    a. 8

    b. 16

    c. 128

    d. 1024

18. Both CRC-16 and CRC-32 error detection transmit data in blocks of _____ bits.

    a. 8

    b. 16

    c. 128

    d. 1024

19. Using _____ ARQ, a sending modem must wait for a return ACK for each sent block before sending the next block.

    a. delivered

    b. efficient

    c. discrete

    d. continuous

20. Using _____ ARQ, a sending modem does not wait for a return ACK for each sent block before sending the next block.

    a. delivered

    b. efficient

    c. discrete

    d. continuous

21. The current ITU standard for analog modem error detection is _____.

    a. odd parity checking

    b. even Parity checking

    c. V.42

    d. V.92

22. The _____ protocol uses a continuous ARQ method.

    a. MNP

    b. NAK

    c. LAP-M

    d. ACK

23. The _____ protocol only sends individual NAK blocks back to a receiving modem.

    a. MNP

    b. NAK

    c. LAP-M

    d. ACK

Section 7.6

24. The V.42bis compression standard does not require V.42 error correction be used. True/False

25. The V.42bis compression standard lists a maximum compression ratio of up to _____ to 1.

    a. 2

    b. 4

    c. 8

    d. 16

26. A realistic compression ratio for the V.42bis compression standard is _____ to 1.

    a. .5

    b. 1.5

    c. 2.5

    d. 4.5

27. Using the V.42bis compression standard, a sending modem transmits a _____ to the receiving modem.

    a. word dictionary

    b. pattern dictionary

    c. word list

    d. pattern template

28. Data compression can always occur, even when two communicating modems are using different compression standards. True/False

29. Using the V.42bis compression standard at a 28.8 kilobits-per-second rate, maximum throughput can be calculated at _____ kilobits per second.

    a. 56

    b. 115.2

    c. 128

    d. 256

30. The latest ITU data compression standard is _____.

    a. V.42

    b. V.42bis

    c. V.44

    d. V.56

Discussion Questions

Section 7.1

1. Sketch the amplitude-shift keying signal representing the binary code of *11000101*.

2. Sketch the amplitude-shift keying signal representing the binary code of *01011010*.

3. Sketch the frequency-shift keying signal representing the binary code of *00111011*.

4. Sketch the frequency-shift keying signal representing the binary code of *11100010*.

5. Sketch the phase-shift keying signal representing the binary code of *10101011*.

6. Sketch the phase-shift keying signal representing the binary code of *11010101*.

7. Sketch a QAM 16 constellation diagram and indicate the location of the two symbols that would represent the binary code of *01011010*.

8. Sketch a QAM 16 constellation diagram and indicate the location of the two symbols that would represent the binary code of *11100010*.

9. Using Nyquist's equation, calculate the maximum channel capacity of a communications system that has a total bandwidth of 5500 hertz.

10. Using Nyquist's equation, calculate the maximum channel capacity of a communications system that has a total bandwidth of 1.1 megahertz.

11. A communications system has a bandwidth of 8 kilohertz. Using QAM 16, find the maximum data rate using Nyquist's equation.

12. A communications system has a bandwidth of 8 kilohertz. Using QAM 64, find the maximum data rate using Nyquist's equation.

Section 7.2

13. A communications system has a signal voltage level of 32 volts$_{rms}$ and signal noise on the system is measured at .45 volt$_{rms}$. Calculate the SNR in decibels.

14. A communications system has a signal voltage level of 25 volts$_{rms}$ and signal noise on the system is measured at .75 volt$_{rms}$. Calculate the SNR in decibels.

15. A communications system has an SNR of 21 decibels. Calculate the *S:N* value.

16. A communications system has an SNR of 43 decibels. Calculate the *S:N* value.

17. A communications system has an *S:N* value of 1500 and a bandwidth of 4000 hertz. Calculate the channel capacity using Shannon's law.

Section 7.3

18. What is the difference between the V.90 and V.92 modulation standards?

Section 7.4

19. The usable bandwidth of a communications system is 6500 hertz. Given an SNR of 30 decibels calculate the channel capacity using Shannon's law.

20. The usable bandwidth of a communications system is 6500 hertz. Given an SNR of 6.5 decibels calculate the channel capacity using Shannon's Law.

Section 7.5

21. Set the parity bit for the large letter G for odd parity.

22. Set the parity bit for the symbol "~" for odd parity.

23. Set the parity bit for the symbol + for even parity.

24. Set the parity bit for the small letter x for odd parity.

25. What is a prime number? Why is it important when using CRC error detection?

26. Describe how discrete ARQ is different from continuous ARQ.

27. Describe how the MNP protocol differs from LAP-M.

Section 7.6

28. Describe how Huffman encoding differs from V.42bis.

29. Research the V.44 data-compression standard. Select a dial-up ISP in your area and determine whether they support this standard.

Broadband for the Masses: Digital Subscriber Line and Cable Modem

Objectives   Upon completion of this chapter, the student should be able to:

- Understand the bandwidth limitations of the existing local loop.
- Describe the effect of loading coils and bridged taps on bandwidth.
- Discuss how, by eliminating loading coils and bridged taps, the bandwidth of the local loop can be extended.
- Comprehend and describe the asymmetry used in ADSL systems.
- Define the frequencies used in ADSL systems and how they are split.
- Describe the difference between full-rate and universal ADSL.
- Understand how a DSL modem functions.
- Describe the function of a DSLAM in a DSL system.
- Understand how peer-to-peer Internet file-sharing applications can cause bottlenecks.
- Describe the ways ADSL signals are modulated.
- Describe other forms of DSL.
- Describe the three major components used in a cable system network.
- Define the cable system channel frequency allocations.
- Describe how the cable system infrastructure has been modified to carry two-way data traffic.
- Understand how cable systems are brought out to neighborhoods.
- Describe the concept of bandwidth sharing on the cable system network.
- Discuss privacy issues on the cable system network.

- Understand the popular modulation methods used with the cable system network.
- Comprehend and describe the asymmetry used in cable system networks.
- Describe the functions of a firewall and a proxy server.
- Contrast cable network and DSL systems, describing the major benefits of each.

Outline

8.1   A Local Loop Review

8.2   Asymmetric Digital Subscriber Line

8.3   Other Forms of Digital Subscriber Line

8.4   Cable System Networks

8.5   Multiple-Device Internet Access Security and Performance

8.6   Is One Access Method Better than the Other?

Key Terms

10BaseT

baseband

carrierless amplitude phase (CAP) modulation

client

community antenna television (CATV)

digital subscriber line (DSL)

digital subscriber line access multiplexer (DSLAM)

discrete multitone (DMT) modulation

distribution point

drop cables

DSL lite

feeder cables

firewall

G.lite

headend

high-bit rate digital subscriber line (HDSL)

high-bit rate digital subscriber line 2 (HDSL2)

hybrid fiber coaxial (HFC)

Internet protocol (IP) address

ISDN digital subscriber line (IDSL)

neighborhood node

network address translation (NAT)

peer-to-peer

proxy server

quadrature amplitude modulation 64 (QAM 64)

quality of service (QoS)

rate-adaptive digital subscriber line (RADSL)

Siamese cable

single-carrier modulation quadrature phase-shift keying (SCM QPSK)

splitter

splitterless DSL

symmetric digital subscriber line (SDSL)

truck roll

universal ADSL (UADSL)

universal serial bus (USB)

very high speed digital subscriber line (VDSL)

Introduction

As Internet bandwidth demand continues to grow, existing voice bandwidth on the local loop is not sufficient. For user satisfaction, high-bandwidth Web applications including E-commerce and streaming audio and video require more bandwidth than the local loop can provide in the legacy state. The telephone companies have discovered broadband technologies like **digital subscriber line (DSL)** and are now using different versions of DSL to extend the life of the local loop copper wire pairs most of us now have coming into our homes and businesses. In addition, the cable television companies have discovered that, with some infrastructure changes, they have provided enough bandwidth to their customers to offer high-bandwidth broadband data and voice along with cable television. Most homes and *small office home offices (SOHOs)* in the United States have cable television access, and the cable companies have upgraded their cabling infrastructure to allow two-way (upstream and downstream) traffic. Cable companies and telephone companies are now competing to sell consumers broadband data services. Today, many cable companies are offering dialtone over their networks and have become cable local-exchange carriers (CLECs) in their respective areas in direct competition with the traditional LECs. Many Internet service providers (ISPs) have decided to support both technologies, and the customer race is on. This chapter initially reviews the limitations of the legacy local loop, discusses how the local loop is changing to accommodate products like asymmetric digital subscriber line (ADSL), and then covers the current *xDSL* offerings. [Each variation, in acronym, ends with the DSL suffix. The prefix is typically different, so "x"DSL is typically used to refer to all variations of digital subscriber line services.] Cable modem will then be covered and compared to xDSL. Finally, the chapter offers a brief look at how local area networks (LANs) are being attached to these common broadband systems.

8.1 A Local Loop Review

The analog public switched telephone network (PSTN) or plain old telephone service (POTS) local loop is defined as the twisted pair of copper wires most people have coming into their homes or businesses. The local loop has been "tuned" to human voice frequencies over the last 100 years and has a bandwidth, including guardbands, of approximately 4000 hertz. Loop resistance is critical in the local loop, and phone companies have had to "tune" the loop to the 300- to 3300 hertz frequency range using loading coils to transmit high-quality voice. Frequencies above 4000 hertz on loaded loops are blocked and, because of the low-pass-filter characteristics of a loaded local loop, a loaded loop cannot provide the high bandwidth required by data services like DSL. As carriers move to provide these high-bandwidth data services, the loading coils must be removed from the local loop circuits.

Not only loading coils but also bridged taps can cause problems when offering high-bandwidth data services, like DSL, on the local loop. A bridged tap is an unterminated wire pair with ends running down the street. Bridged taps can create an impairment to the transmission system. A signal on the loop moves down the unterminated cable and will reflect back to the main pair affecting the main signal. A bridged tap will typically not be noticed at voice transmission frequencies; but, at the higher frequencies required by high-bandwidth data services like DSL bridged taps may cause a significant problem and typically must be removed.

It is estimated that there are 200 million copper local loops in the United States. In addition to the loading coils and bridged taps, a single local loop in the United States has an average of 19 splices with each splice producing its own signal reflection and echo.

8.2 Asymmetric Digital Subscriber Line

There are several variations of *digital subscriber line (DSL),* a telecommunications service that allows high-bandwidth transmission to homes and businesses over the existing local loop telephone lines; the current variations are described in this chapter. *Asymmetric digital subscriber line (ADSL)* is currently the most popular offering from LECs across the United States. *Asymmetric digital subscriber line (ADSL)* is a variation of DSL that allows the phone companies to expand the usable bandwidth of the existing local copper loop by removing the loading coils and bridged taps. By removing the loading coils and bridged taps from local loops, the bandwidth of the local loop is extended from approximately 4 kilohertz to 1.1 megahertz on local loops up to 3.6 kilometers, as illustrated in figure 8-1.

FIGURE 8-1
Local loop
bandwidth.

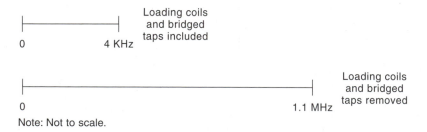

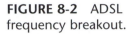

FIGURE 8-2 ADSL frequency breakout.

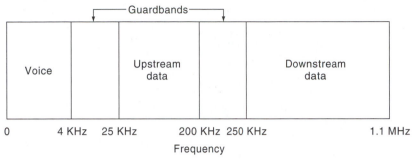

Note: Diagram not to scale.

The term *asymmetric* in ADSL refers to the service's differing *upstream* (away from the user) and *downstream* (toward the user) bandwidths. The ADSL data service is considered an *always-on* technology; once connected, the service is always available. High-bandwidth data is transmitted simultaneously with circuit-switched voice conversation using *frequency division multiplexing (FDM),* a multiplexing technique that uses different frequencies to represent different communications channels. The ADSL FDM frequency breakout is shown in figure 8-2.

Separating the Voice Frequencies

With ADSL, voice calls use the normal 0- to 4-kilohertz spectrum and the DSL modem uses the higher frequencies to pass data traffic. Many phones generate frequencies higher than 4 kilohertz, which can interfere with the data stream. In addition, the higher DSL frequencies can be picked up by the phone and cause headset static. Let us look at how the separation is done and how the noise problem has been solved.

Full-Rate ADSL

The original solution to the 4-kilohertz interference problem was to use frequency **splitters.** A splitter is a filtering device that is attached to the phone line and splits the line; one branch hooks up to the original house telephone wiring, and the other branch heads to the DSL modem. In the telephone company central office, another splitter splits the voice and data. Home and central office splitters are illustrated in figure 8-3.

Besides splitting the phone line, the splitter acts as a low-pass filter, allowing only 0- to 4-kilohertz frequencies to pass to and from the phone, and eliminates the 4-kilohertz interference between phones and DSL modems.

With early modulation techniques, full-rate ADSL systems required the installer to go to the customer location and configure the ADSL modem to a fixed bit rate in both the upstream and downstream directions. The bit rates, once set, could not be changed unless a technician came out to the customer location again and made the changes. Understanding that local

FIGURE 8-3
Full-rate ADSL.

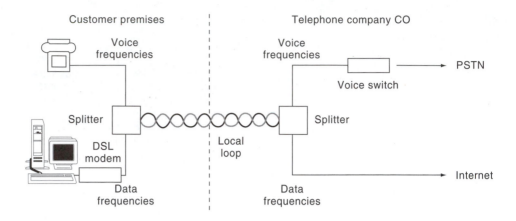

loop conditions can change significantly with variables like moisture and temperature, setting the bit rates on a single day with unique weather conditions made little sense.

Universal ADSL and G.lite

Customer site splitter installation requires a **truck roll,** which is the industry term used to indicate a telecommunications technician visit to a customer site. When setting up ADSL access, truck rolls typically require that new wiring be installed to the DSL modem and can get expensive. **Universal ADSL (UADSL)** is a variation of ADSL that was devised by the *Universal ADSL Working Group (UAWG)* in 1997 to solve the truck roll problem by allowing the customer to install the premises' equipment. The UAWG consists of most **Regional Bell Operating Companies (RBOCs)** and many large and small companies including Microsoft and Intel. Both the *International Telecommunications Union–Telecommunication Standardization Sector (ITU-T)* and the *ADSL Forum* have supported universal ADSL and have selected the **G.lite** term to refer to the UADSL standard.

G.lite, also commonly referred to as **DSL lite** or **splitterless DSL,** is an ADSL compromise that takes the changing local loop conditions into consideration by lowering bandwidth. G.lite runs at 1.5 megabits per second downstream and up to 512 kilobits per second upstream with asynchronous transfer mode (ATM) used as transport protocol. With G.lite, the splitting is being done only at the telephone company CO; the technology is referred to as *splitterless.* When both voice and data are passed on the same loop, low-pass *microfilters* must be placed on each telephone to separate the high-frequency data and low-frequency voice signals. This is illustrated in figure 8-4. G.lite, when originally proposed, appeared to be the direction this technology was going but recently has fallen out of favor to **rate-adaptive digital subscriber line (RADSL)** due to G.lite's low bandwidth limitations.

FIGURE 8-4
DSL lite.

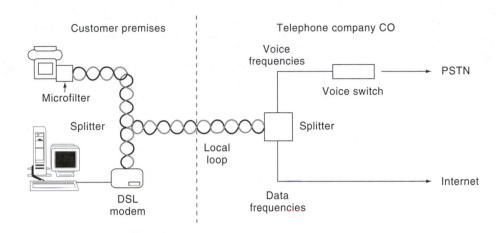

Rate-Adaptive Digital Subscriber Line

Both full-rate DSL and G.lite technologies require a fixed data rate, with full-rate DSL offering varying data rates that must be technician set and G.lite offering only a single upstream and downstream rate. *Rate-adaptive DSL (RADSL)* is a DSL variation that, in concept, allows the DSL modems to dynamically change with varying line conditions. At this time RADSL rate changes only occur when the modem is powered off and on and, once the data rate is set on power up, the modem stays at this data rate until power is recycled again. RADSL allows bit rates of between 1.5 and 8 megabits per second downstream and 1.544 megabits per second upstream over local loop distances of up to 1.5 kilometers depending on bandwidth. Most ADSL equipment currently being installed is actually RADSL, and the terms *ADSL* and *RADSL* have become synonymous.

DSL Modem

The DSL modem is considered to be *customer premises equipment (CPE)*. The modem is commonly a two-port device with an *RJ-11* connector for attachment for the local loop and an Ethernet **10BaseT** *RJ-45* connector for PC local area network (LAN) connection, as illustrated in figure 8-5.

In addition to a 10BaseT Ethernet connector, many ADSL modems optionally have a **universal serial bus (USB)** for connection to the PC. Ethernet is the most commonly used local area network (LAN) technology in use today at bit rates up to 1 gigabit per second. To use Ethernet requires that an Ethernet network interface card be installed in the device being attached. Universal serial buses (USBs) are commonly built into newer PCs and are device ports that support a data bandwidth of 12 megabits per second.

The DSL modem modulates and demodulates the respective signals using one of two modulation methods described later in this chapter. Ethernet is

FIGURE 8-5 DSL modem connection detail.

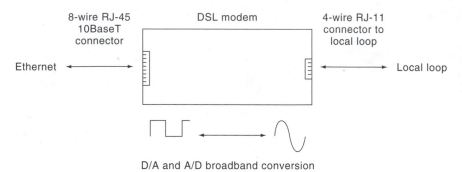

D/A and A/D broadband conversion

Note: Connectors not to scale.

considered **baseband** digital signaling. A *baseband* communications system is a system that transmits information on a single channel. This is in contrast to the local loop connection, which is considered *broadband* analog signaling. A broadband communications system is a system that transmits information on multiple channels simultaneously.

Digital Subscriber Line Access Multiplexer (DSLAM)

A **digital subscriber line access multiplexer** (**DSLAM)** is a frequency division multiplexer that is used in the LEC CO to multiplex multiple customer data connections onto a high-bandwidth Internet connection. The DSLAM controls and routes DSL data traffic between the customer premises equipment and the network service provider's network. A major advantage of DSL and DSLAMs is the fact that data is moved off the voice switch, freeing up the more expensive voice switch technology to handle what it was originally designed for, voice traffic. The DSLAM is linked to the Internet via a router commonly using asynchronous transfer mode (ATM), as illustrated in figure 8-6.

Recall that asynchronous transfer mode (ATM) is a connection-oriented digital switching technique that dynamically allocates bandwidth using

FIGURE 8-6 DSLAM/ATM router link.

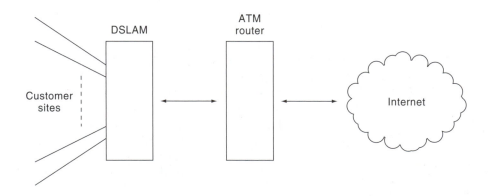

FIGURE 8-7 ADSL data asymmetry.

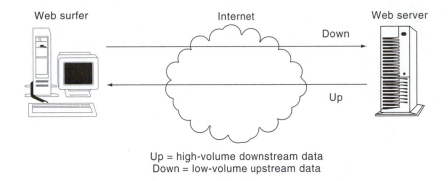

Up = high-volume downstream data
Down = low-volume upstream data

53-byte, fixed-size packets called *cells*. It is highly efficient and offers several advantages for data traffic, including a high **quality of service (QoS).** Quality of service (QoS) is an indication of a system's ability to provide a service and is very important when transferring streaming audio and video.

ADSL Data Rate Asymmetry and Bandwidth

The difference in upstream and downstream data bandwidth is significant, and ADSL has been designed for the Internet Web surfer. When using basic Internet services like E-mail, Web browsing, and streaming audio and video, the vast majority of traffic is being delivered to the user in the downstream direction. Upstream traffic is typically used only for requests for data to be delivered in the downstream direction, as illustrated in figure 8-7. From the figure, it is easy to see that ADSL has been designed for the home user/Web surfer. ADSL has not been designed to offer Web services to others on the Internet, such as setting up a Web server, because these types of services require higher-volume upstream data bandwidth that ADSL does not offer but other variations of xDSL do. These are discussed later in the chapter.

Peer-to-Peer Networks

With the growing popularity of **peer-to-peer** Internet-sharing applications like Napster (http://www.napster.com) and Gnutella (http://gnutella. wego.com) ADSL users may experience some upstream bandwidth bottlenecks. Peer-to-peer networks are simple network configurations that allow individual users to administer and control network access to their machines. Sharing applications like Napster and Gnutella are peer-to-peer programs that users install on their machines. Let us use Napster as an example. Users go to the Napster Web site and download the free sharing program. Users then install this program on their machines. The program allows all users in the world currently attached to the Internet and running Napster to share their files with all other users who are attached to the Internet and running Napster. Once a user downloads a file, that user can then share that same

FIGURE 8-8 Peer-to-peer traffic flow.

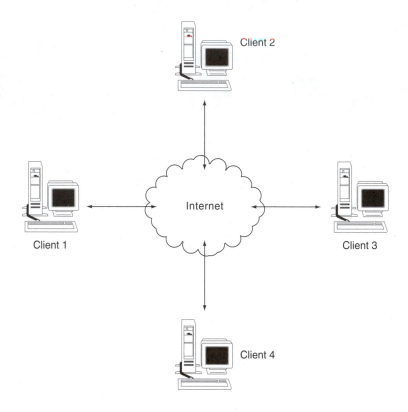

downloaded file with all other Napster users. Dedicated Web servers are not required for users to share files. The Napster program that users download and install on their machines makes users' machines, running desktop operating systems like Windows 2000 or Windows XP, function as peer-to-peer servers on the Internet, as illustrated in figure 8-8. In the figure, the upstream bandwidth constraints are apparent. Each **client** on a network is an individual attached computer. Let us say that client 1 is sharing a music file that clients 2, 3, and 4 want but do not have. If clients 2, 3, and 4 see that client 1 is sharing the file and each decides to download the file from client 1, then client 1's upstream bandwidth will be split among the three requesting clients, resulting in a slow downstream download to each of the three. This bottlenecking is illustrated in figure 8-9.

ADSL Signal Modulation

Two signal modulation methods are currently used with ADSL: **discrete multitone (DMT) modulation** and **carrierless amplitude phase (CAP) modulation.**

FIGURE 8-9 Peer-to-peer upstream sharing bandwidth limitation.

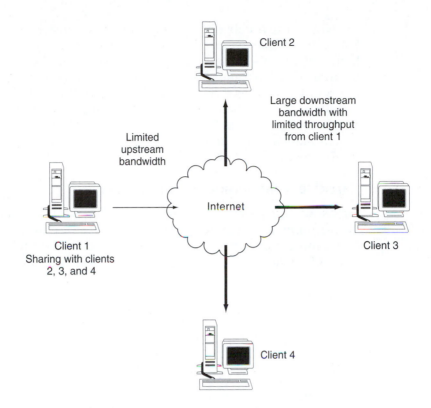

Discrete Multitone Modulation

Discrete multitone (DMT) modulation is a modulation technique that takes the upstream and downstream frequency ranges and separates them into 256 frequency bands of 4.3125 kilohertz each. Each channel uses the same QAM used by analog modems described in chapter 7. QAM levels from 2 to 8 (4 to 256 constellation points) are typical for each channel. Common channel assignments are listed in table 8-1.

TABLE 8-1 Common DMT channel assignments.

Channels	Use	Direction	Bandwidth range
1	Voice	Bidirectional	0–4 KHz
2–5	Guardband		4 KHz–24 KHz
6–31	Data	Upstream	24 KHz–134 KHz
32–43	Guardband		134 KHz–186 KHz
44–250	Data	Downstream	186 KHz–1.1 MHz

Notice the number of upstream versus downstream channels in table 8-1. Downstream bandwidth is much higher than the upstream bandwidth—hence, the term *asymmetric*. By using a large number of channels and transmitting a variable number of bits on each channel, DMT can segment regions of the frequency bandwidth that have low signal-to-noise ratios. DMT has the support of both Intel and Microsoft; it also has an *American National Standards Institute* (*ANSI:* http://www.ansi.org) and a *European Telecommunications Standards Institute* (*ETSI:* http://www.etsi.org) standard; and it is currently the method of choice.

Carrierless Amplitude Phase Modulation

Carrierless amplitude phase (CAP) modulation is a proprietary DSL modulation standard implemented by *Globespan Semiconductor* (http://www.globespan.net). CAP modulation is different from DMT modulation in that it uses all of the available bandwidth as a single channel and then optimizes the bit rate to the existing line conditions. CAP channel assignments are shown in table 8-2. The different downstream baud rates allow the CAP modem to adapt to varying line conditions.

TABLE 8-2
Common CAP
channel assignments.

Direction	Carrier frequency	Baud rate
Upstream	113.2 KHz	136 Kbaud
Downstream	435.5 KHz	340 Kbaud
Downstream	631 KHz	680 Kbaud
Downstream	787.5 KHz	952 Kbaud

CAP uses a modulation technique referred to as *two-dimensional QAM.* Using two-dimensional QAM, CAP splits a full-rate bit stream into two half-rate bit streams. These half-rate streams are then individually modulated by two digital filters. The modulation carriers in each stream are then canceled by each other, and the combined bit stream is sent out over the local loop to the LEC CO. DMT and CAP systems offer comparable downstream and upstream bit rates. DMT offers better performance, utilizes bandwidth more efficiently, and consumes less power. For this reason, DMT has become the modulation method of choice for ADSL.

8.3 Other Forms of Digital Subscriber Line

Several other digital subscriber line (DSL) solutions are currently in various stages of development. All DSL solutions support simultaneous POTS except

IDSL. Some are briefly described in the following paragraphs with each assuming 24-gauge wire is used when referring to distances.

Very High Speed Digital Subscriber Line

Very high speed digital subscriber line (VDSL) is a DSL variation that provides between 13 and 52 megabits per second downstream and between 1.6 and 2.3 megabits per second upstream over distances of up to 1.5 kilometers. It is being explored primarily as a means to bring video-on-demand services to the home.

High Bit Rate Digital Subscriber Line

High bit rate digital subscriber line (HDSL) is a DSL variation that is symmetrical in that both the upstream and downstream bit rates are equivalent. It is considered a T-1 "killer" and is a business-marketed product. HDSL uses two or three pairs of wire offering 1.544 megabits per second over two copper pairs and 2.048 megabits per second over three pairs. Local loop distance is restricted to 1.2 kilometers, but phone companies can install signal repeaters to extend this range up to 12 kilometers.

Symmetric Digital Subscriber Line

Symmetric digital subscriber line (SDSL) is a DSL variation that also offers upstream and downstream channels of equal bandwidth on a single wire pair. Upstream and downstream bandwidth is offered at 768 kilobits per second on local loop distances of up to 4 kilometers. Most SDSL systems use 2B1Q digital modulation, while some use CAP.

High Bit Rate Digital Subscriber Line 2

High bit rate digital subscriber line 2 (HDSL2) is a DSL variation touted as the *next-generation HDSL* within the American National Standards Institute (ANSI: http://www.ansi.org) and the European Telecommunications Standards Institute (ETSI: http://www.etsi.org). It offers symmetric 2 megabits per second upstream and downstream over a single pair and uses pulse amplitude modulation (PAM). HDSL2 became a standard on May 21, 2001.

ISDN Digital Subscriber Line

ISDN digital subscriber line (IDSL) is a data-only DSL service that provides full-duplex-data throughput at bandwidths up to 144 kilobits per second. IDSL uses the same 2B1Q modulation code as ISDN to deliver service without special line conditioning but, unlike ISDN, is a nonswitched service, so it does not cause switch congestion at the service provider's CO. IDSL is also considered an always-on service and does not require the call setup that ISDN requires.

8.4 Cable System Networks

Early cable systems were typically in rural areas and consisted of large antennas on mountains with coaxial cable run from the antennas to local subscribers. These original systems were referred to as **community antenna television (CATV).** In the early 1970s cable satellites were launched and cable companies began receiving these satellite signals and offering many channels to their customers. It was at this point that the cable companies moved from the more rural areas into the city suburbs and began to compete with the local television companies offering wireless access. Today, cable television is widespread in the United States with approximately 63 percent of the population subscribing to some cable service and 95 percent of the U.S. residences having cable access (Source: http://www.javaworld.com/javaworld/jw-10-1996/jw-10-connors.html).

The cable network was originally designed and constructed for the downstream delivery of television signals only and consists of three major components, the cable **headend,** cable **trunks,** and cable **distribution points.**

Cable Headend

The cable system *headend* is the location where all cable system signals are received for delivery on the local cable network. The headend typically consists of one or more satellite links and a building that houses transmission equipment. At the headend, programs are assigned channels and moved onto the cable network for downstream delivery to customers.

Cable Trunks

Cable *trunks* are cables that carry the cable signals away from the headend toward the customers. Coaxial cable trunk systems typically include signal conditioners/amplifiers every 2000 to 3000 feet.

Cable Distribution Points

Cable *distribution points* are devices that move the television signals off the trunk cable to smaller **feeder cables** and are typically used to serve a neighborhood. Feeder cables are cables that are typically run through residential neighborhoods and are tapped with **drop cables,** which are the cables that

FIGURE 8-10 The cable television network.

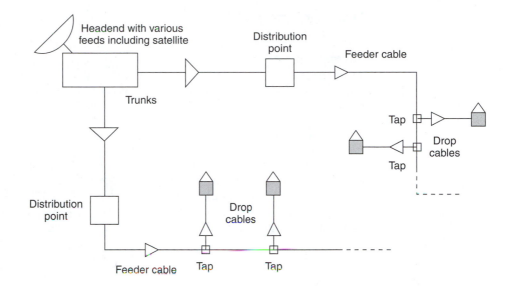

enter the customer premises. The cable television network distribution system is illustrated in figure 8-10.

CATV Frequency Allocation

Original cable systems consisting of only coaxial cable provide a bandwidth of either 330 or 450 megahertz and this bandwidth is used to provide video to downstream subscribers. The system designers had the foresight in their original design to allow some frequency space for future upstream traffic. The first downstream channel, channel 2, occupies the space between 55.25 and 61.25 megahertz for both video and audio transmission. Each sequential channel occupies 6 megahertz of available bandwidth, as shown in figure 8-11.

A 450-megahertz traditional coaxial cable CATV system that uses the frequency allocation outlined in figure 8-11 can carry a total of 60 analog television channels. Most cable systems now use a **hybrid fiber coaxial (HFC)** cabling system. A hybrid fiber coaxial (HFC) cabling system is partially fiber and partially coaxial cable with the fiber handling higher-bandwidth trunking and the coaxial handling the lower-bandwidth connections. Using an HFC system expands the available subscriber bandwidth to 700 megahertz and allows the delivery of 110 channels.

FIGURE 8-11 Channel frequency allocation.

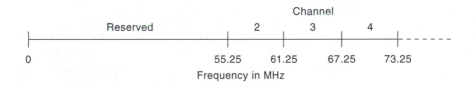

Data on the Cable

Cable television networks were originally constructed for the delivery of downstream traffic only. To provide the necessary upstream bandwidth, cable providers have two basic options. Upstream bandwidth can be provided over the voice telephone network, or the cable television network infrastructure can be modified for two-way traffic.

Upstream with the PSTN

Upstream bandwidth can be provided over the public switched telephone network (PSTN), while downstream bandwidth is provided over the installed cable infrastructure, as illustrated in figure 8-12. This was an initial but temporary solution used by many cable companies that wanted to provide data to their customers.

FIGURE 8-12
Upstream cable data over the PSTN.

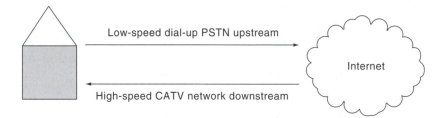

Cable Television Infrastructure Modification

Today, most cable companies that have moved into the data business have modified their networks to support simultaneous upstream and downstream transmission. These television/data networks are typically HFC with fiber used for trunking and feeding while coaxial cable is still typically used to bring the lines into a home or business, as illustrated in figure 8-13.

FIGURE 8-13
Typical CATV data network.

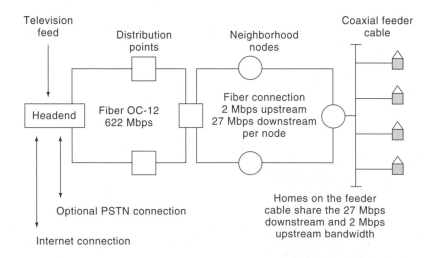

FIGURE 8-14
Siamese cable.

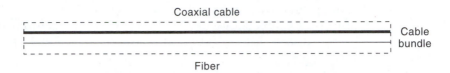

A modern headend serves 100,000 to 300,000 customers, and distribution points will server 10,000 to 30,000 customers. A **neighborhood node** serves local neighborhoods and can serve up to 1000 customers. Cable companies typically run **Siamese cable** on all feeder lines. Shown in figure 8-14, Siamese cable is cable that includes both coaxial and fiber cable in the same bundle. The coaxial cable is initially used, with the fiber going unused. By running Siamese cable, the cable companies can easily extend the neighborhood node network out to new subdivisions and homes as they are built without running new fiber cable.

Bandwidth Sharing

Both upstream and downstream access to the Internet are shared by all users on a neighborhood node, and this is a selling point that DSL providers have been using when competing with the cable companies. Let us look at a bandwidth example using shared access. Let us say that you are the first customer on your node to purchase Internet access from your local cable provider. As the first user, you have access to all of the upstream and downstream bandwidth *and* Internet access is extremely fast.

Now look at the opposite extreme. Let us say that your neighborhood node is saturated with 1000 users. Now let us say that all users on the neighborhood node are surfing the Internet at the same time. The upstream and downstream access is now shared among 1000 users. We can do some simple throughput calculations for each user in this situation:

$$\text{Upstream bit rate} = \frac{2 \text{ Mbps}}{1000 \text{ customers}} = 2000 \text{ bps}$$

$$\text{Downstream bit rate} = \frac{27 \text{ Mbps}}{1000 \text{ customers}} = 27,000 \text{ bps}$$

These are extreme conditions, but it can be seen that the available bandwidth to each individual user under these conditions is miniscule. Cable companies monitor traffic throughput and are constantly adjusting their networks so that these types of saturation conditions rarely happen.

Privacy

In addition to bandwidth-sharing performance issues, privacy issues must be considered. Cable feeder lines are shared, which means that all customers on the shared line have the ability to see the other feeder line customer's traffic using protocol analyzers and other tools available on the Internet. Customers must be careful about what they share on any machine attached to the feeder lines, because others on the same line may be able to get access to it.

Cable Data Signal Modulation

Many cable companies are now offering individual customers access up to 30 megabits per second downstream and 768 kilobits per second upstream. The two most popular modulation techniques are **single carrier modulation quadrature phase-shift keying (SCM QPSK)** and **quadtrature amplitude modulation 64 (QAM 64).**

Single-Carrier Modulation Quadrature Phase-Shift Keying

Single-carrier modulation quadrature phase-shift keying (SCM QPSK) is a cable modulation technique that generates one of four possible constant amplitude states, with each separated by 90 degrees. SCM QPSK is also referred to as *distributed queue dual bus (DQDB)* and provides downstream bandwidth of up to 10 megabits per second. Two bits per baud are transmitted, producing constellation patterns shown in figure 8-15.

FIGURE 8-15 QPSK constellation pattern.

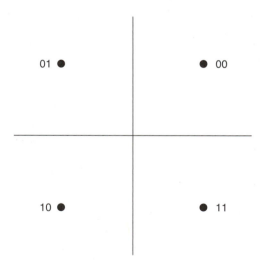

Quadrature Amplitude Modulation 64

Quadrature amplitude modulation 64 (QAM 64) is a cable modulation technique that provides downstream bandwidth of up to 36 megabits per second with eight bits per baud transmitted. Quadrature amplitude modulation was covered in detail in chapter 7 of this text.

CATV Upstream Data Bandwidth

Upstream bandwidth is narrow and limited to between 5 and 40 megahertz. This range is very noise sensitive and picks up interference from poor cabling and bad terminations. In a neighborhood, the media is shared and all this noise gets added together as the signals travel upstream, combining and increasing. Due to this problem, many cable system manufacturers are still using SCM QPSK for modulation even though it offers less bandwidth than QAM 64. SCM QPSK is more efficient in a noisy environment, and the reduced bandwidth has not caused a problem yet. Both ADSL and cable modems use 10BaseT Ethernet as an interface to the PC, and this limits the connection speed to 10 megabits per second or less even in situations in which the modem can actually receive at 30 megabits per second.

8.5 Multiple-Device Internet Access Security and Performance

As the number of computer systems in homes and businesses grows, the need for simultaneous Internet connectivity by multiple devices grows along with it. Many homes and small businesses now have local area networks (LANs) installed for printer and file sharing. It no longer makes sense to put a modem in each Internet-capable device and have them dial out individually. Several products are available now that allow multiple devices to share a single Internet connection simultaneously. In fact, Microsoft Windows 98 Second Edition added Internet-connection sharing that allows multiple devices to share an analog modem, xDSL, cable modem, or Ethernet connection to the Internet. This feature has continued to be offered in all newer versions of Windows. Many of our homes and businesses are now always connected to the Internet using always on-technologies like xDSL and cable modem. With connectivity come security issues and performance issues. **Firewalls** and **proxy servers** are common ways to deal with these issues.

Firewalls and Proxy Servers

Customers with more than one computer needing Internet access are now installing firewalls and/or proxy servers that provide security and allow a single analog modem, ADSL, or cable modem connection to be shared among multiple machines.

Firewalls

A *firewall* is a device used to prevent unauthorized access to a specified network. All devices connected to the Internet using the TCP/IP are assigned an

FIGURE 8-16
Typical firewall/proxy
server configuration.

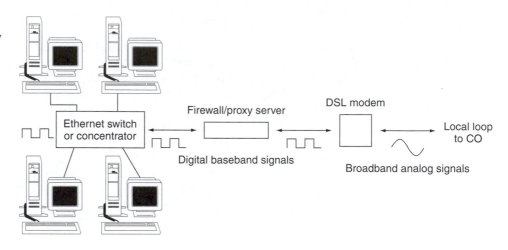

address called an **Internet protocol (IP) address.** An Internet protocol (IP) address is a unique 32-bit binary address assigned to any network device attached to a network that uses transmission control protocol/Internet protocol (TCP/IP). Firewalls typically use an Internet protocol addressing technique called **network address translation (NAT)** on all connected devices behind the firewall inside the customer site. This is illustrated in figure 8-16. This addressing technique provides internal IP addresses to all PCs on the internal network. These internal addresses cannot be seen by others on the Internet who may attempt to hack into the customer machines. The only address that is seen to the rest of the Internet is the address of the firewall device, as illustrated in figure 8-17.

Without the firewall, sensitive information is relatively easy to get to by experienced hackers. This scenario is shown in figure 8-18. In the figure, PC-1 may have sensitive data that the customer does not want anyone else to access. If PC-1 was connected directly to the cable modem, PC-1 would have a valid IP address and others on the Internet could potentially hack into the machine and get access to the sensitive data the customer wants to protect. Firewalls provide a level of security that will prevent many Internet hackers from stealing, modifying, or destroying data on machines behind the firewalls.

Proxy Servers

A *proxy server* is a device that is typically combined with a firewall and is used to store Web sites as they are accessed by users behind the firewall. Proxy servers increase the performance when Web surfing. For example, if a proxy server is in use and a user goes to a particular site on the Internet, that site is downloaded from the remote Web server, viewed by the user, and also

FIGURE 8-17 NAT addressing firewall/proxy server configuration.

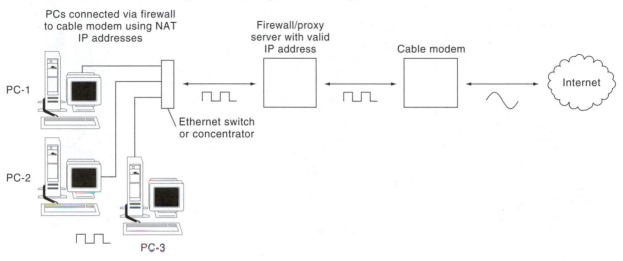

FIGURE 8-18
PC directly
connected to DSL
modem.

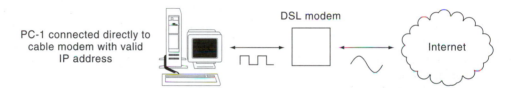

cached, or stored, on the proxy server. If another user then goes to the same site, the system does not have to reach all the way out to the remote Web server to download the site; the site has been cached on the proxy server, and the user only needs to go as far as the proxy server to retrieve the site. An example is shown in figure 8-19. The first request for the site by PC-1 requires that the actual site be accessed and the site be downloaded. The site data, on the way to PC-1, passes through the proxy server and is cached, or stored, on the proxy server.

Further along in the day, let us say that PC-3 requests the same site. When this happens, PC-3 does not actually have to go out to the remote site to retrieve the Web site; it only needs to go as far as the proxy server, as illustrated in figure 8-20.

Proxy servers allow the administrator to set refresh-rate levels, and also typically allow site restrictions, and sometimes offer added features like virus scanning.

FIGURE 8-19 Proxy server
Example: first request by PC-1
for http://www.nctt.org

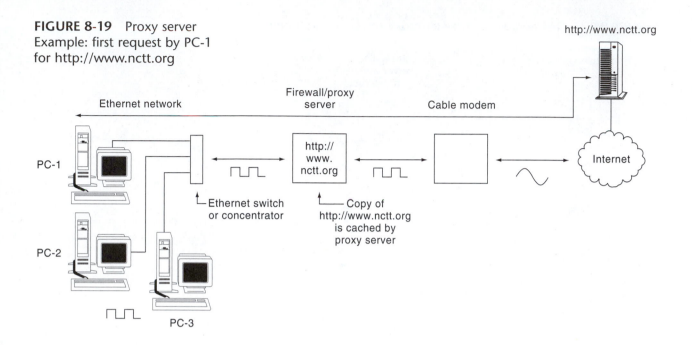

FIGURE 8-20 Proxy server
example: second request by
PC-3 for http://www.nctt.org

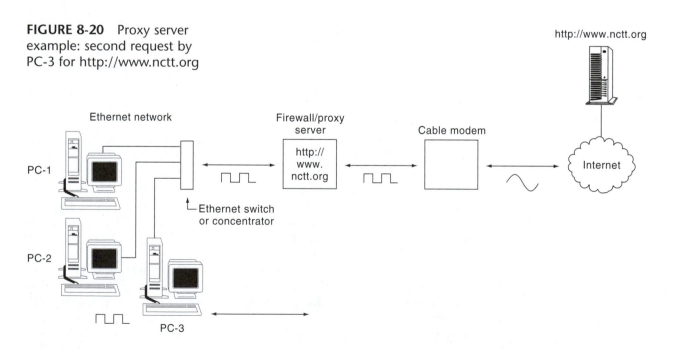

8.6 Is One Access Method Better than the Other?

Several studies have compared and contrasted ADSL and cable modem. Keynote Systems has put together a very nice independent study contrasting ADSL, cable modem, and T-1 (http://www.keynote.com/measures/dsl). ADSL has experienced growing pains, and the LECs have had problems with widespread implementation. Many blame the poor overall condition of the local loop and its inability to carry high-frequency DSL signals. Others feel that the indecision over G.lite has caused the LECs to be set back in the race.

Regardless of the technology used for access, the Internet continues to grow and customer bandwidth expectations grow right along with it. Many experts feel that bandwidth-hungry applications such as streaming video, audio, and IP telephony are growing so rapidly that backbone providers may not be able to keep up. According to Mark Lottor of Network WIZARDS (http://www.nw.com), who is running the longest Internet domain survey, as of July 2002, the Internet size had reached over 162 million hosts; this is significantly more than the 43 million hosts identified in June 1998. With the demand for Internet backbone bandwidth increasing fast and with technologies like WDM solving the long-haul bandwidth problem, the local access bandwidth gap quickly is being closed with technologies like xDSL and cable modem.

Summary

1. The existing *local loop* has been tuned to voice frequencies limiting bandwidth.

2. *Loading coils* and *bridged taps* must be removed from the local loop to increase loop bandwidth from approximately 4 kilohertz to 1.1 megahertz.

3. *ADSL* is currently the most popular broadband data offering from LECs in the United States.

4. Voice and data frequencies are *split* in ADSL systems.

5. *UADSL* systems require a truck roll to a customer site for splitter installation.

6. *G.lite* (also known as *DSL lite* or *splitterless DSL)* and *RADSL* do not require a truck roll.

Low-pass microfilters are installed on each telephone to separate the lower-voice frequencies and the higher-data frequencies.

7. A *DSL modem* is a two-port device with an RJ-11 connector for local loop attachment and either an Ethernet or USB for computer connection.

8. DSL modems *modulate* and *demodulate* respective signals.

9. A *DSLAM* is used in the LEC CO to split or separate voice and data frequencies.

10. The DSLAM is a bank of *DSL modems* that are linked to the Internet via a *router* commonly using *asynchronous transfer mode (ATM)*.

11. ATM has a high *quality of service (QoS)*.

12. ADSL systems are *asymmetric* with more bandwidth allocated in the *downstream* direction.

13. Peer-to-peer network applications like Napster and Gnutella can cause *upstream bottlenecks* in ADSL systems.

14. Two modulations are currently used with ADSL: *DMT* and *CAP*.

15. Both DMT and CAP use *QAM* modulation techniques.

16. Other forms of DSL include *VDSL, HDSL, SDSL, HDSL2,* and *IDSL*.

17. The *cable network* was originally designed for downstream delivery of television signals.

18. *Sixty-three percent* of U.S. residences subscribe to some cable service, and *95 percent* of U.S. residences have cable access.

19. The three major components of the cable system network are the *cable headend, cable trunks,* and *cable distribution points*.

20. A *single-cable channel* occupies *6 megahertz* of bandwidth.

21. *Hybrid fiber coaxial (HFC)* systems expand the available cable bandwidth to *700 megahertz* and allow delivery of up to *110 different television channels*.

22. The *cable television network infrastructure* has been modified to support simultaneous upstream and downstream traffic.

23. Cable network bandwidth is *shared* among users at the neighborhood level.

24. It is possible to view other users' traffic on *cable feeder lines*.

25. Two modulation methods are currently used with cable data: *SCM QPSK* and *QAM 64*.

26. *Firewalls* use *NAT addressing* and provide a level of security from Internet hackers.

27. A *proxy server* is typically combined with a firewall and increases Web surfing performance by storing visited Web sites.

28. Both *DSL* and *cable modem* systems have advantages and disadvantages over each other.

Review Questions

Section 8.1

1. _____ and _____ limit the bandwidth of the telephone system local loop. (Choose two.)
   a. CO batteries
   b. Loading coils
   c. Telephone handsets
   d. Bridged taps

2. The voice-tuned local loop has a bandwidth of approximately _____ .
   a. 300 Hz
   b. 3300 Hz
   c. 4 KHz
   d. 1.1 MHz

Section 8.2

3. A local loop that has been modified to carry ADSL data traffic along with voice has an approximate bandwidth of _____ .
   a. 300 Hz
   b. 3300 Hz
   c. 4 KHz
   d. 1.1 MHz

4. The term _____ refers to ADSL's differing upstream and downstream bandwidths.
   a. asymmetric
   b. digital
   c. subscriber
   d. line

5. Splitters are used in ADSL systems to separate voice and data _____.

   a. amplitudes

   b. frequencies

   c. phases

   d. telephones

6. _____ requires a technician to come out and install a splitter device.

   a. G.lite

   b. Universal ADSL

   c. Full-Rate ADSL

   d. DSL lite

7. _____ offers varying data rates that are set by telephone company technicians.

   a. G.lite

   b. Universal ADSL

   c. RADSL

   d. DSL lite

8. A DSL modem typically has a/an _____ input and a/an _____ output. (Choose one for each blank.)

   a. RJ-11

   b. Arcnet

   c. Token Ring

   d. Ethernet

9. A/An _____ is a CO device that functions as a frequency division multiplexer separating the voice and data frequencies.

   a. ATM router

   b. switch

   c. SLC-96

   d. DSLAM

10. The benefits of _____ include dynamic bandwidth allocation using cells, high efficiency, and high quality of service (QoS).

    a. ATM

    b. ASK

    c. FSK

    d. ADSL

11. The majority of common Internet surfing traffic moves in the upstream direction from the surfing customer. True/False

12. Peer-to-peer networks do not require dedicated Web servers. True/False

13. Napster is an example of a _____ networking application.

    a. client-server

    b. text-based

    c. peer-to-peer

    d. downstream only

14. _____ network applications can cause upstream bandwidth bottlenecks.

    a. Client-server

    b. text-based

    c. peer-to-peer

    d. downstream only

15. _____ ADSL signal modulation takes the upstream and downstream frequency ranges and separates them into 256 frequency bands of 4.3125 kilohertz each.

    a. PSK

    b. DMT

    c. ASK

    d. CAP

16. _____ ADSL signal modulation is a proprietary standard implemented by Globespan Semiconductor.

    a. PSK

    b. DMT

    c. ASK

    d. CAP

Section 8.3

17. _____ provides between 13 and 52 megabits per second downstream and between 1.6 and 2.3 megabits per second upstream over distances of up to 1.5 kilometers, depending on bandwidth.

   a. RADSL

   b. SDSL

   c. VDSL

   d. IDSL

18. _____ offers 1.544 megabits per second over two wire pairs and 2.048 megabits per second over three wire pairs.

   a. HDSL

   b. SDSL

   c. IDSL

   d. VDSL

19. _____ offers a bandwidth of 768 kilobits per second in the upstream and downstream directions over a single wire pair.

   a. HDSL2

   b. SDSL

   c. IDSL

   d. VDSL

20. _____ offers 2 megabits per second upstream and downstream over a single pair.

   a. HDSL2

   b. SDSL

   c. IDSL

   d. VDSL

21. _____ provides an upstream and downstream bandwidth of 144 kilobits per second.

   a. HDSL2

   b. SDSL

   c. IDSL

   d. VDSL

Section 8.4

22. The cable _____ in the cable network system is the point where all cable system signals are received for delivery on the local cable network.

   a. distribution point

   b. trunk

   c. modem

   d. headend

23. A cable _____ in the cable network system carries the signal away from the point where all cable system signals are received for delivery on the local cable network.

   a. distribution point

   b. trunk

   c. modem

   d. headend

24. The cable _____ in the cable network system is used to serve a typical neighborhood and move signals off larger cables to smaller feeder cables.

   a. distribution point

   b. trunk

   c. modem

   d. headend

25. Each television channel on the cable network occupies _____ of available bandwidth.

   a. 4000 Hz

   b. 9600 Hz

   c. 1.1 MHz

   d. 6 MHz

26. Cable television networks were originally designed to deliver signals in the downstream (to the customer) direction. True/False

27. Most cable systems now use a/an _____ cabling system that uses a combination of coaxial cable and optical fiber.
   a. ASK
   b. CAP
   c. DMT
   d. HFC

28. A typical cable system neighborhood node can serve up to _____ customers.
   a. 1000
   b. 5000
   c. 30,000
   d. 300,000

29. A typical cable system headend can serve up to _____ customers.
   a. 1000
   b. 5000
   c. 30,000
   d. 300,000

30. A typical cable system distribution point can serve up to _____ customers.
   a. 1000
   b. 5000
   c. 30,000
   d. 300,000

31. _____ cable includes both coaxial and fiber cable.
   a. UTP
   b. Siamese
   c. STP
   d. POTS

32. Cable modem systems typically share bandwidth with other neighborhood users. True/False

33. This cable data signal modulation method—_____—provides downstream bandwidth of up to 10 megabits per second and is also known as DQDB.
   a. DMT
   b. QAM 64
   c. SCM QPSK
   d. CAP

34. This cable data signal modulation method—_____—provides downstream bandwidth of up to 36 megabits per second.
   a. DMT
   b. QAM 64
   c. SCM QPSK
   d. CAP

35. Both ADSL and cable modems typically use _____ as an interface to the PC.
   a. Token Ring
   b. Arcnet
   c. ATM
   d. Ethernet

Section 8.5

36. Firewalls typically use an IP addressing technique called _____ on all connections inside the firewall.
   a. TCP
   b. MAC
   c. NAT
   d. BIOS

37. _____ are commonly combined with a firewall and are used to store commonly visited Web sites.
   a. Proxy servers
   b. Client servers
   c. Web servers
   d. Stand-alone servers

Discussion Questions

Section 8.1

1. Explain why the voice-frequency-tuned local loop does not provide sufficient bandwidth for DSL services.

2. Do bridged taps affect DSL service as much as loading coils? Why or why not? In your answer, explain the difference between a bridged tap and a loading coil.

Section 8.2

3. Explain why ADSL is considered an *asymmetric* service.

4. What must be done to separate the ADSL voice and data frequencies? What is used to do this?

5. Describe the differences between universal ADSL and G.lite.

6. Call your local telephone provider and determine if DSL is available in your area. Can you get the service on your phone line?

7. Describe the difference between a baseband and a broadband system. Give at least one example of each.

8. What is used in the CO to separate voice and data frequencies? In your answer, be sure to explain where the voice and data frequencies go after separation.

9. Explain how a peer-to-peer application like Gnutella can significantly increase ADSL upstream bandwidth and create a bottleneck.

10. Briefly describe the two forms of DSL modulation commonly used. Is one more popular than the other? Why or why not?

Section 8.3

11. Make a chart and compare upstream bandwidth, downstream bandwidth and distance for VDSL, HDSL, SDSL, HDSL2, and IDSL.

12. Your aunt wishes to run a Web server out of her home office for her growing business. What form of DSL would you recommend that she have installed? Why?

13. Is HDSL2 available in your area? Contact your LEC and any CLECs and find out. How much does it cost? When calling, also ask about T-1 availability and its cost. Is there a difference in HDSL2 and T-1 cost?

Section 8.4

14. Explain why early cable television systems were not capable of providing both upstream and downstream bandwidth over the cable. What was an early solution used to provide upstream bandwidth?

15. Do you have cable modem availability in your neighborhood? If so, what is the cost per month?

16. Explain why having a large number of cable modem users on your cable network will decrease data performance.

17. Will an application like Gnutella produce the same upstream bandwidth problems on both ADSL and cable modem connections? Explain.

18. Is the cable data network considered asymmetric? Explain.

19. Describe the two methods of data modulation used by the cable network. Is one more popular than the other? Explain.

Section 8.5

20. What is the function of a firewall? Is there one on your campus?

21. What is the function of a proxy server?

Section 8.6

22. Contact your cable company and telephone provider and create a table contrasting cable modem and ADSL in your area. In the table, include upstream bandwidth, downstream bandwidth, and monthly cost.

23. Visit http://www.nw.com. What is the current Internet domain host count?

Computer Networks

Objectives Upon completion of this chapter, the student should be able to:

- Define local, campus, metropolitan, and wide area networks.
- Discuss the differences between a peer-to-peer and a server-based network.
- Describe the difference between read-only and full network access.
- Identify peer-to-peer network security issues.
- Realize the advantages of server-based networks.
- Describe local-area-network bus topology and why it is typically not used today.
- Identify and explain bridged taps.
- Discuss the benefits of star local-area-network topology.
- Discuss local-area-network ring topology.
- Describe the OSI model and discuss how it works at each of the seven layers.

Outline 9.1   Network Configurations
9.2   Network Designations
9.3   Network Topologies: Connecting the Computers
9.4   The Open System Interconnect Model

Key Terms

application layer
bus topology
campus-area network (CAN)
collapsed ring

communications protocol
data-link layer
encryption
enterprise network

full access
Internet Engineering Task Force (IETF)
logical link control (LLC) sublayer

media access control (MAC) sublayer

metropolitan-area network (MAN)

network administrator

network interface card (NIC)

network layer

open system interconnect (OSI) model

physical layer

presentation layer

read-only access

request for comments (RFC)

ring network

server-based network

session layer

software driver

star topology

terminating resistance

transport layer

wide-area network (WAN)

Introduction

Most computer users today are no longer working on stand-alone machines. In the office, computers and other devices like printers are usually connected to each other using a local-area network (LAN). In the home and in small businesses, analog modems, ADSL, and cable modems are common ways that can be used to access the Internet, check E-mail, and securely connect to a remote office LAN. In chapter 8, we saw that both ADSL and cable modem provide high-bandwidth access to the Internet and that both ADSL and cable modems allow computer network connectivity and sharing via an Ethernet port. In this chapter, we take a look at some fundamentals necessary to understand computer networking.

9.1 Network Configurations

Networks are typically connected in local, campus, metropolitan, and wide-area network configurations.

Local-Area Networks

Local-area networks (LANs) are the simplest networks and consist of a collection of computers and peripherals in a small area connected together to share resources. Figure 9-1 shows a typical small office/home office (SOHO) LAN configuration.

Ethernet is commonly used today for device connectivity, and each individual device must have an Ethernet network card installed to connect into

FIGURE 9-1 Typical small office/home office (SOHO) local-area network.

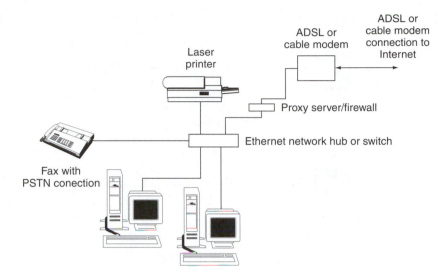

the hub or switch. The Internet connection, printer, and fax machine can be shared by all computers on the network. In addition, each computer can be shared with the other computers on the network. The proxy server/firewall is used primarily for security and also content caching. Firewalls protect internal LANs from external attacks by hackers on the Internet.

Campus-Area Networks

Many organizations have a campuslike network configuration that allows computer users in different buildings to share information and resources. **Campus-area networks (CANs)** are computer networks that are used to interconnect LANs in buildings that are in close proximity to each other, as shown in figure 9-2. CAN connectivity can be done with a variety of methods. Bandwidth can range from relatively low values of 10 megabits per second up to 1 gigabit per second.

Metropolitan-Area and Wide-Area Networks

Usually, as a business grows, employees, home offices, branch offices, manufacturing facilities, and so on are added. These new facilities may be across town, across the state, or across the world. **Metropolitan-area networks (MANs)** and **wide-area networks (WANs)** are networks used to connect remote LANs into what is typically called an **enterprise network,** which is a network that connects multiple LANs that are geographically distributed within an organization. (See figure 9-3 and figure 9-4 for examples of a MAN and a WAN.)

FIGURE 9-2 Typical campus-area-network configuration.

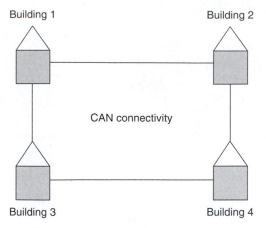

FIGURE 9-3 Metropolitan-area-network example.

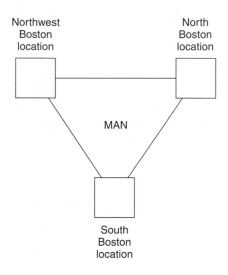

FIGURE 9-4 Wide-area-network example.

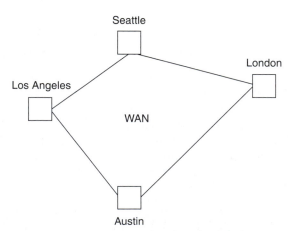

9.2 Network Designations

The two fundamental computer network designations we discuss in this chapter are **peer-to-peer** and **server-based networks.**

Peer-to-Peer Networks

Peer-to-peer computer networks are computer networks that allow all users on the network to share resources. For example, a single printer, fax machine, or modem connection can be shared in a peer-to-peer networking environment. In the sample peer-to-peer network shown in figure 9-5, the contents of each PC's hard disk drive and devices like printers and scanners can be shared with all other PCs on the network, saving the time and expense of duplicating equipment and transferring data using floppy, Zip, or CD-ROM drives. Microsoft has offered built-in peer-to-peer, local-area networking in all versions of Windows since the Windows for Workgroups version 3.11 operating system was released in November 1992. Figure 9-6 provides an illustration of the Windows shared-folder indicator.

FIGURE 9-5 Sample peer-to-peer network.

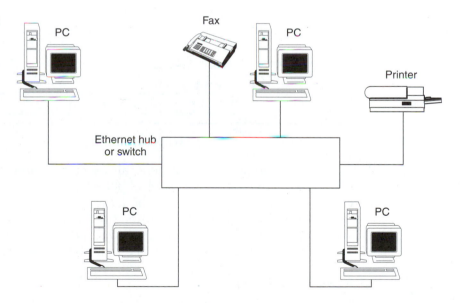

FIGURE 9-6 Microsoft shared-folder (hand) indicator.

Sample

FIGURE 9-7 Sample properties dialogue box showing folder-sharing options with Windows XP Professional.

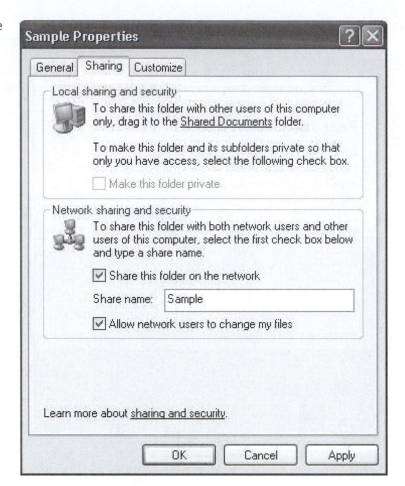

In a peer-to-peer environment, all users are responsible for allowing access and sharing of folders and attached devices on their individual machines. Passwords optionally can be set on each share, and security is referred to as *share level*. Figure 9-7 shows the Windows XP Professional folder-sharing options. There is no centralized administration for a peer-to-peer network, and folder and device access can be set as *read-only* or *full*.

Read-Only Access

If a user wants to share a folder on his or her machine, the user typically has the option of setting the folder as *read-only*. **Read-only access** is a network sharing option that only allows others attached to the network to read files, as indicated in figure 9-8. In the figure, PC-1 is sharing the folder C:\Sample. The other PCs on the network can see the Sample folder and its contents on PC-1 and can open and/or download any files in the folder. However, with

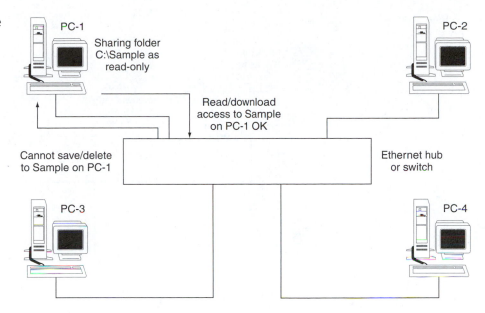

FIGURE 9-8 Sample read-only access, peer-to-peer network.

read-only access, PC-2, PC-3, and PC-4 cannot save any information to the Sample folder on PC-1.

Full Access

The user on PC-1 also has the option of giving **full access** to users. Full access is another network sharing option that allows others attached to the network to read, download, save, and delete files in the shared folder.

Figure 9-9 indicates that, by giving full access to the Sample folder on PC-1, users on PC-2, PC-3, and PC-4 will be able to read, download, save, and delete in the Sample folder on PC-1.

Passwords

The user has the option in a peer-to-peer environment to set different passwords on individual shares, and shares can be set up with no password. Most peer-to-peer operating systems allow passwords to be set for read-only and/or full access.

In figure 9-10, the Windows ME Sample folder has been set for "depends-on-password" access. The user on PC-1 will select and enter separate passwords for read-only and full access. This password must then be communicated (by voice, E-mail, etc.) to other users on the network.

As indicated in figure 9-11, users on other machines on the network who know passwords will be able to access shared folders (read-only or full) only if they know the proper password. If a share password is set and is not known, then the user will not have any access.

FIGURE 9-9 Sample full-access, peer-to-peer network.

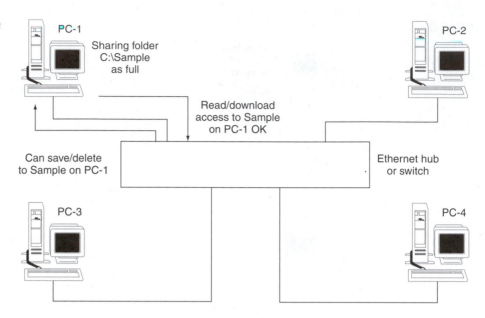

FIGURE 9-10 Sample folder dialogue box set for depends-on-password access with Windows ME.

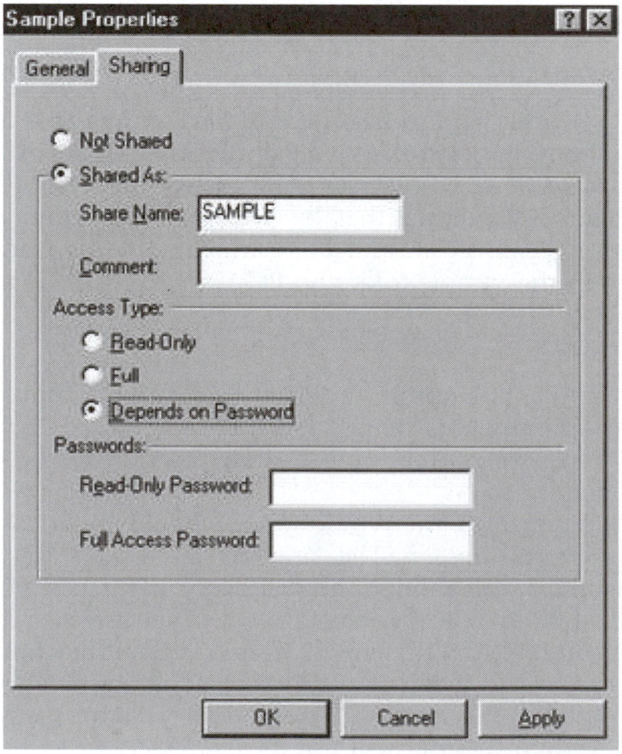

FIGURE 9-11
Sample depends-on-password access, peer-to-peer network.

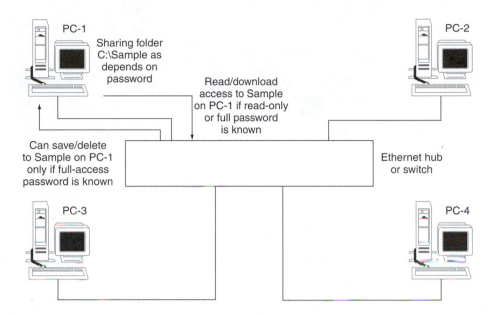

Peer-to-Peer Network Security

Peer-to-peer network security is extremely weak. Passwords are set at the share level with all users accessing a share having to know exactly the same password. When password access is set up this way, passwords have a tendency to leak to unwanted users. It is also an inconvenience for users to change password access on their machines because any new passwords must then be given to all authorized users. For these reasons, peer-to-peer networks are only recommended for small numbers of users; operating system manufacturers typically recommend peer-to-peer networking for 10 users or less. **Server-based networks** are recommended for larger numbers of users.

Server-Based Networks

Server-based networks (figure 9-12) are networks that are designed for larger groups of users than peer-to-peer networks. In a server-based-network environment, different servers are typically dedicated to different functions. For example, servers can be dedicated to things like file storage, printing, faxing, E-mail, backup, and storage.

Servers are typically higher-end machines running advanced operation systems like Windows 2000 Server, Novell NetWare, or Linux. These operating systems allow centralized user level access administration and can support thousands of individual users. Server-based networks are managed by a designated **network administrator.** The network administrator is the

FIGURE 9-12
Sample server-based
network.

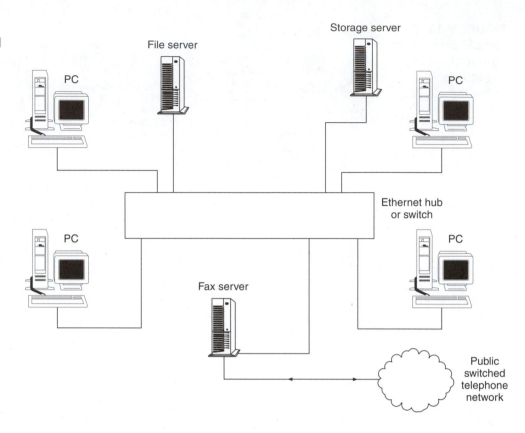

person who has total control of the server-based network and controls the
access of all users with each user set up with a unique username and password. Access to different folders and devices on the network is also controlled by the administrator. For example, if an administrator does not want
a specific user to have access to a specific folder on a file server, the administrator can turn off access to that user and grant access to other users that
require access, as illustrated in figure 9-13. In the figure, notice that Mary
and Nina have been granted and have access to the Sample folder to the network file server. Doug and Tom have not been granted access and cannot see
or access the Sample folder on the file server.

Server-Based Network Security

Server-based networks are much more secure than peer-to-peer networks. Ad-
ministrators can easily turn off access for a single user—for example, when
an employee leaves the company or moves to another department within
the company—and maintain security without having to change a share-level
password and then distribute the new password to all users. The administrator only needs to go into the user administration application for the server
and disable access for the individual user.

FIGURE 9-13
Server-based-network user access example.

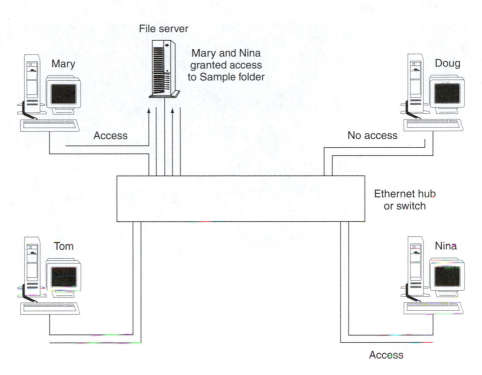

9.3 Network Topologies: Connecting the Computers

Computers can be physically connected in a bus, star, ring, or mesh configuration. Each type of physical connection, or topology, has advantages and disadvantages. In a computer, a **network-interface card (NIC),** as shown in figure 9-14, is the device used to physically connect the computer to the network transmission medium. Once the NIC has been placed in the computer, **software drivers** for the card are installed. Software drivers are programs that allow the operating system to access and use the NIC to attach to the network. Once software drivers are installed, **communications protocols** are loaded and configured. Communications protocols are pieces of software that set communications rules that allow two devices to talk to each other. For two connected network devices to communicate with each other, they both must be running the same communications protocol.

Once the card is installed, software drivers and protocols are loaded, and the card is physically attached to the network (figure 9-15), a device can access and use the network.

FIGURE 9-14
Ethernet 10/100-megabits-per-second network-interface card.

FIGURE 9-15
Network interface card installation and configuration steps.

Bus Topology

A **bus topology** attaches all devices (PCs, printers, etc.) to a single cable that is commonly referred to as the network *backbone* or *trunk,* as indicated in figure 9-16. A **terminating resistance** on each cable end is required for cable termination to prevent transmitted signals from bouncing back onto the network and mixing with other transmitted signals.

Bus topology networks are typically not used any more because they are unreliable and difficult to troubleshoot. In a bus topology network, if a single connector opens or short-circuits, the entire network goes down. The most common bus topology network is *10Base2* and is covered in the next chapter.

Star Topology

A **star topology** network uses individual cable attached to a centralized connection hub for each device attached to the network. The most commonly used network access method, *Ethernet,* is typically configured in a physical star topology while it actually operates using a logical bus topology.

FIGURE 9-16
Bus topology.

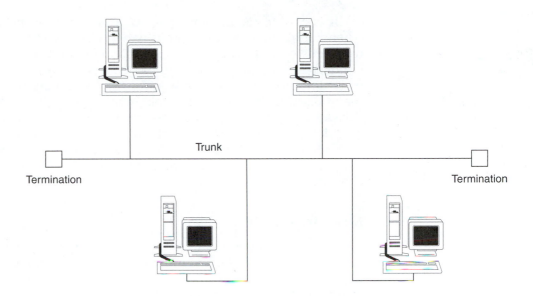

Physical Star

Ethernet networks today are commonly designed and constructed in a physical star configuration using two different types of centralized connection devices, *concentrators* (also commonly referred to as *hubs*) and *switches*. Each Ethernet device to be connected to the network has an Ethernet NIC installed and an individual twisted-pair cable is run from the device to the centralized concentrator or switch, as illustrated in figure 9-17.

FIGURE 9-17
Physical star
topology.

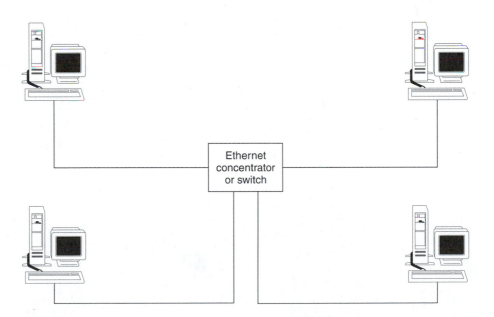

It is easy to see the advantage of a physical star network configuration when compared to a physical bus network configuration. If a single connector opens or short-circuits in a star configuration, only the network device connected with that cable/connector loses network access; the entire network does not go down as it would in a physical bus topology network. The major disadvantage of a star topology is the centralized connection device. If this device goes down (power loss, lightening strike, etc.), the entire network goes down. The most common star topology networks in use today are Ethernet *10BaseT* and *100BaseT*. These networks, which use unshielded-twisted-pair (UTP) copper wire to make the connections, are covered in the next chapter.

Ring Topology

In addition to bus and star topologies, local-area networks can be configured in a ring configuration. A **ring network** is a network constructed similar to a bus network with cable run from network device to network device. Instead of terminating the ends, as in a bus network, the two ends are joined, as illustrated in figure 9-18.

The actual ring does not have to be formed using cable run from network device to network device; it can be in the form of a **collapsed ring.** A

FIGURE 9-18
Ring topology.

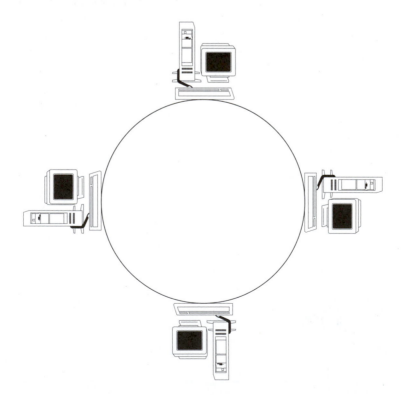

FIGURE 9-19
Collapsed-ring
topology.

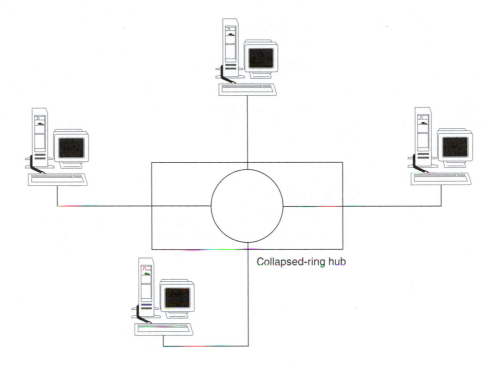

Collapsed-ring hub

collapsed-ring network is a ring network with the ring formed inside a connection hub. The cable topology resembles a star network, as illustrated in figure 9-19. The most common ring topology in use today is *Token Ring,* which is covered in the next chapter.

9.4 The Open System Interconnect Model

The **open system interconnect (OSI) model** is a network structure for standard interfaces and communication protocols developed and promoted by the International Standards Organization (ISO: http://www.iso.ch). The OSI model was designed as a generic model that works with all network physical topologies, logical topologies, and communications protocols. Communications protocols are the sets of rules that devices in a telecommunications network use to communicate with each other. The OSI model uses seven logical layers to define data communications between network devices:

7. Application
6. Presentation
5. Session

4. Transport

3. Network

2. Data Link

1. Physical

The layers are listed in reverse order (7–1) for clarity and understanding. This section begins with a basic overview; then, each layer is discussed individually.

Data Transfer

Let us first take a look at how information moves in a network from the application layer on one machine to the application layer on another machine. We do not have to start at the application layer of the OSI model but will do so here to provide a complete overview.

When we use the term *application* and reference it to the *application layer* of the OSI model, we are referencing *network* applications and not *program* applications like Microsoft Word or Excel. For example, most of us are familiar with E-mail and get numerous messages each day. The actual program applications each of us uses to create, send, receive, and read E-mail may be different but we can all communicate with each other. Some of the more popular E-mail application programs include Microsoft Outlook, Lotus Notes, Eudora, and Novell GroupWise. The user interface is different with each of these E-mail applications, but they each perform the same fundamental functions and are compatible. If they were not compatible, we would need to use only Outlook to open messages created in Outlook, only Eudora to open messages created with Eudora, and so on. This scenario would make it a hassle to read E-mail from different people, and we all realize we do not have to do this. The reason we do not have to do this is, in part, because of a protocol called *simple mail transfer protocol (SMTP)*. The SMTP is a protocol that runs at the application layer of the OSI model. The specific E-mail program that one uses does not run at the application layer of the OSI model, but the SMTP does and, combined with other pieces of the E-mail program, allows different program applications (such as Outlook or Eudora) to read messages created with other program applications, as illustrated in figure 9-20. The SMTP is one of several application layer protocols and is only being used as an example here.

Now that we have differentiated between a program application and a network application, let us look at how an E-mail message moves from one user to another over the Internet, as shown in figure 9-21. The message starts at the sending computer that is running an E-mail program such as GroupWise or Outlook. When the user clicks the send button in the E-mail program, the message is eventually launched into the network in the form of a series of *packets* of *1*s and *0*s. Before an E-mail message packet can get out onto the Internet, it must pass through each of the seven layers of the

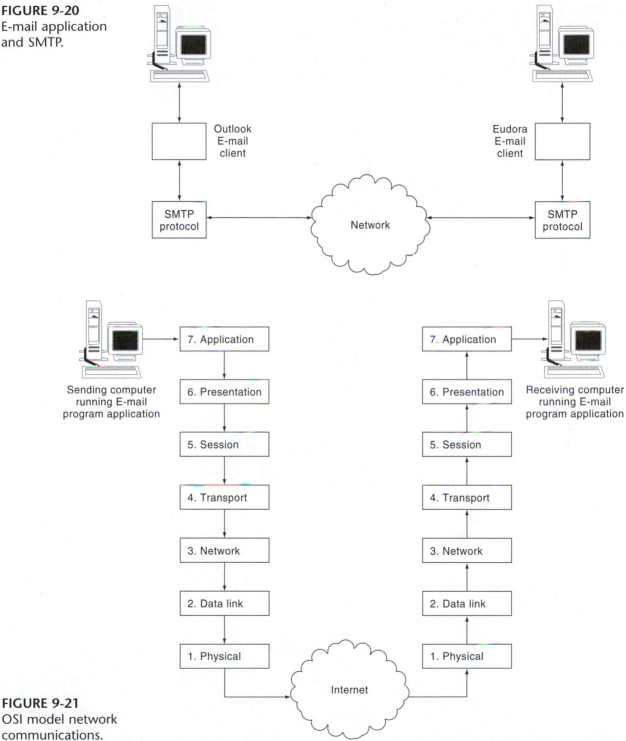

FIGURE 9-20
E-mail application and SMTP.

Outlook E-mail client

Eudora E-mail client

SMTP protocol

Network

SMTP protocol

Sending computer running E-mail program application

7. Application

6. Presentation

5. Session

4. Transport

3. Network

2. Data link

1. Physical

7. Application

6. Presentation

5. Session

4. Transport

3. Network

2. Data link

1. Physical

Receiving computer running E-mail program application

Internet

FIGURE 9-21
OSI model network communications.

FIGURE 9-22
OSI model network layer-3-to-layer 3 communications.

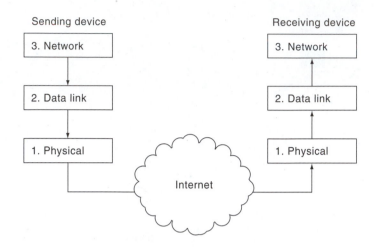

OSI model. Each layer adds bits to the packet, with each layer responsible for a different function (addressing, error correction, compression, encryption, etc.) required to move the packet from sender to receiver. Once the packet is finished in the *physical layer,* it has been encapsulated with each layer's additional bits, commonly referred to as *overhead,* and is now ready for transmission over the Internet to the receiving device. The packet finds its way from the sending computer to the receiving computer and comes in on the receiving side at the physical layer. The packet then moves up the OSI model on the receiving computer, and bits that were added on the sending computer are stripped off at the corresponding layer at the receiving computer. This results in a message that can be read with any common E-mail program on the receiving computer.

For successful end-to-end communications over a network, packets must pass sequentially through layers of the OSI model. However, depending on what kind of information is being transmitted, every packet does not have to start at the application layer. For example, transmission can start at the sending device at layer 3 and packets can pass from layer 3 on the sending device to layer 3 on the receiving device, as illustrated in figure 9-22. In this example, the upper layers 4 through 7 are not involved.

When two machines are communicating over a network, each OSI model layer on each computer acts like it is talking to its associated layer on the other computer. In addition, each lower layer provides required services to the next higher layer and these services are considered "transparent" to the higher layers. The next higher layer has no concern regarding the lower layer information so there is no need for the higher layers to "see" what the lower layers are doing. For example, the *network layer* provides address information to the *transport layer.* The transport layer has no concern with network layer addressing; it is only concerned with transport layer protocol and function. Network-layer addressing is handled by the network layer only. Each layer

FIGURE 9-23
OSI model apparent versus actual communications.

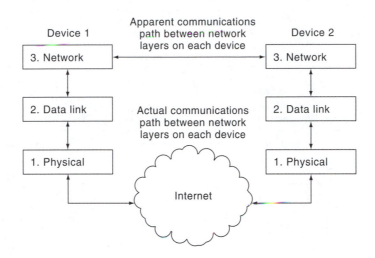

does not "see" what the other layers are doing. Two devices communicating network layer to network layer are illustrated in figure 9-23. In the figure, at each network device, the communication appears to be between the network layer on the sending device and the network layer on the receiving device, as indicated in green. In actuality, any transfer of data from network layer to network layer requires all data to pass from the network layer on the sending device to the sending-device *data-link layer,* to the sending-device physical layer, through the transmission media to the receiving-device physical layer, up through the receiving-device-data link layer to the receiving device network layer.

It is critical that all information pass through each lower layer successfully. If one layer fails in its function, communication fails. This is indicated in figure 9-24. In the figure, the data-link layer on device 1 has failed in its function (the data-link layer's function is discussed later in this chapter). As a result, communication between the network layers on each device has failed.

The Internet Engineering Task Force and Request for Comments

Let us next look at how communications standards are set. With the basic OSI model explanation just discussed, one can easily see that complete and detailed sets of rules are required for devices to communicate across a network. These sets of rules are called *protocols;* let us now look at how these protocols are defined.

The **Internet Engineering Task Force (IETF:** http://www.ietf.org) is an international organization that controls how communications protocols are set up. The IETF is open to anyone interested in the Internet and is divided

FIGURE 9-24
OSI model
communications
error.

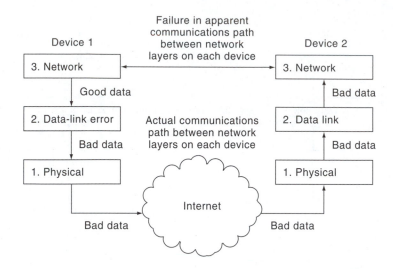

into working groups, which are divided by topic. The IETF, as a group, holds three meetings a year, but most of the work is done using E-mail lists. Each working group has a mailing list, and there are individual lists for general discussion and also for announcements. Each working group also maintains a **request for comments (RFC).** A request for comments (RFC) is a document that includes definitions of Internet policy and protocol. All Internet communications protocols are based on RFCs. The RFCs define each protocol and how each one is used. The first RFC was published by Steven Crocker on April 7, 1969, when the Internet was known as the *ARPANET.* ARPANET had been theorized by Paul Baran and was designed as a distributed redundant communications network that could survive a major enemy attack. A copy of the first RFC network working group request for comment:1 can be found at ftp://ftp.isi.edu/in-notes/rfc1.txt

Baran's original memorandum from 1964 can be found at: http://www.rand.org/publications/RM/RM3420

In addition, historical ARPANET maps can be found at http://www.cybergeography.org/atlas/historical.html

The RFCs are used to keep track of and maintain notes on a particular topic. Currently, over 3000 RFCs exist, with the complex index listed and hyperlinked at http://www.faqs.org/rfcs/rfc-index.html

These RFCs range from notes (see RFC 128: http://www.faqs.org/rfcs/rfc129.html) to protocol development (see RFC 2024: http://www.faqs.org/rfcs/rfc2024.html) to security issues (see RFC 3090: http://www.faqs.org/rfcs/rfc3090.html). Reviewing the list provides an interesting historical view of the development of the Internet. Each RFC has an assigned editor who reviews all input before publishing to the specific document. For more information, to see archived general discussion and announcement list messages and to subscribe to a list, see: http://www.ietf.org/maillist.html

Also see *The Tao of IETF* at: http://www.ietf.org/tao.html for a good historical overview of the IETF. Following is a detailed review of each layer of the OSI model.

Layer 7: Application Layer

As noted earlier, the term *application,* when used at the **application layer,** references network applications like the simple mail transfer protocol (SMTP) and not program applications like Microsoft Word or Excel. The application layer of the OSI model has two major functions; it provides a common interface to program applications, and it determines if the required resources are available for communications to occur. Also noted earlier was how the SMTP works at the application layer. The common application layer protocols are the World Wide Web (WWW), file-transfer protocol (FTP), Telnet, and the simple mail transfer protocol (SMTP).

Layer 6: Presentation Layer

The **presentation layer** of the OSI model controls the conversion of information from the application layer. It is responsible for encoding information from the application layer into specific formats that the receiving device can decode back to the original information. The presentation layer is also responsible for **data compression** and **encryption.** *Data compression* reduces the number of bits that need to be transmitted in a communications session and saves time and money. *Encryption* is the process of scrambling data so that only authorized receivers can unscramble the data. Some common presentation-layer standards include Joint Photographic Experts Group (JPEG), Tagged Image File Format (TIFF), Motion Picture Experts Group (MPEG), and the American Standard Code for Information Interchange (ASCII).

Layer 5: Session Layer

The **session layer** of the OSI model is responsible for establishing and maintaining communications sessions. These communications are coordinated between protocols at this layer. For example, the session layer manages dialogue control. In a *full-duplex environment,* where network traffic can move in two directions at the same time, the session layer is the layer that allows traffic to go in both directions simultaneously. In a *half-duplex environment,* where traffic cannot move in both directions simultaneously, the session layer will only allow traffic to move in one direction at a time and will control which device can communicate. Also, in the event of a transport-layer error, the session layer will initiate the re-creation of the transport-layer connection. Session-layer protocols include Appletalk Session Protocol, NetBEUI, XWindows, Network File System (NFS), Structured Query Language (SQL), and Remote Procedure Call (RPC).

Layer 4: Transport Layer

The **transport layer** of the OSI model provides reliable end-to-end communications and is responsible for ensuring that a communications session is established and then torn down, or removed, when the session is complete. The transport layer also hides the details of the network from the higher layers. The two most common transport-layer protocols are Universal Datagram Protocol (UDP) and Transmission Control Protocol (TCP).

Layer 3: Network Layer

The **network layer** of the OSI model provides the method for addressing and routing data when two different networks are connected together. The network layer is independent from the actual data communications technology (Ethernet, Token Ring, etc.) being used. It routes data from one node to another using logical addressing as opposed to physical addressing. The devices that interpret logical addresses and control data movement are called *routers*. Some common network-layer protocols are the Internet Protocol (IP), Novell's Internetwork Packet Exchange (IPX), and Apple's Datagram Delivery Protocol (DDP).

Layer 2: Data-Link Layer

The **data-link layer** of the OSI model provides reliable error-free delivery of data over the network transmission medium. The specific topology—commonly Ethernet or Token Ring—is also specified at the data-link layer, and the data is framed according to the topology used. The data-link layer is split up into two sublayers, the **media access control (MAC) sublayer** and the **logical link control (LLC) sublayer.**

Media Access Control Sublayer

The *media access control (MAC) sublayer* is the lower of the two data-link sublayers and is responsible for the physical connection sharing among computers connected to the network. Each computer has its own unique MAC address, which is the computer's unique hardware number used to communicate on the local network.

Logical Link Control Sublayer

The *Logical Link Control (LLC) sublayer* is the upper of the two data-link sublayers and identifies the specific line protocol, such as IBM's Synchronous Data-Link Control (SDLC) and Novell's NetWare or NetBIOS. The LLC can also assign sequence numbers to frames and track receiver acknowledgments coming back to the sender.

Layer 1: Physical Layer

The **physical layer** of the OSI model is where bits are put on and taken off the specified transmission media, which can be copper, fiber, or wireless. Specifications at the physical layer include connectors, data rates, voltage levels, and voltage changes. Multiport repeaters, amplifiers, and transceivers are physical-layer devices.

Summary

1. *Local-area networks (LANs)* consist of a collection of computers and peripherals in a small area connected together to share resources.

2. *Campus-area networks (CANs)* are computer networks that are used to interconnect LANs in buildings that are in close proximity to each other.

3. *Metropolitan-area networks* (MANs) and *wide-area networks (WANs)* are networks used to connect remote LANs into what is typically called an enterprise network.

4. An *enterprise network* is a network that connects multiple LANs that are geographically distributed within an organization.

5. *Peer-to-peer* computer networks are computer networks that allow all users on the network to share resources.

6. *Read-only access* is a network sharing option that only allows others attached to the network to read files.

7. *Full access* is a network sharing option that allows others attached to the network to read, download, save and delete files in the shared folder.

8. *Server-based networks* are networks that are designed for larger groups of users than peer-to-peer networks.

9. A *network interface card (NIC)* is used to physically connect a computer to the network.

10. *Software drivers* are programs that allow an operating system to access and use the NIC to attach to a network.

11. *Communications protocols* are pieces of software that set communications rules that allow two devices to talk to each other on a network.

12. A *bus topology* attaches all devices (PCs, printers, etc.) to a single cable that is commonly referred to as the network *backbone* or *trunk*.

13. A *star topology* network uses individual cable attached to a centralized connection hub for each device attached to the network.

14. A *ring network* is a network constructed in a ring topology with backbone ends joined together.

15. The *open system interconnect (OSI) model* is a network structure for standard interfaces and communication protocols developed and promoted by the International Standards Organization.

16. The *Internet Engineering Task Force (IETF)* is an international organization that controls how communications protocols are set up.

17. A *Request for comments (RFC)* is a document that includes definitions of Internet policy and protocol.

18. The *application layer* of the OSI model has two major functions; it provides a common

interface to program applications, and it determines if the required resources are available for communications to occur.

19. The *presentation layer* of the OSI model controls the conversion of information from the application layer.

20. The *session layer* of the OSI model is responsible for establishing and maintaining communications sessions.

21. The *transport layer* of the OSI model provides reliable end-to-end communications and is responsible for ensuring that a communications session is established and then torn down, or removed, when the session is complete.

22. The *network layer* of the OSI model provides the method for addressing and routing data when two different networks are connected together.

23. The *data-link layer* of the OSI model provides reliable error-free delivery of data over the network transmission medium.

24. The *physical layer* of the OSI model is where bits are put on and taken off the specified transmission media.

Review Questions

Section 9.1

1. _____ are network configurations consisting of a collection of connected devices typically in one room or floor of a building.
   a. Local-area networks
   b. Campus-area networks
   c. Metropolitan-area networks
   d. Wide-area networks

2. _____ are network configurations used to connect remote LANs between cities. Select each correct answer.
   a. Local-area networks
   b. Campus-area networks
   c. Metropolitan-area networks
   d. Wide-area networks

3. _____ are network configurations used to connect multiple buildings that are located in the same building complex.
   a. Local-area networks
   b. Campus-area networks
   c. Metropolitan-area networks
   d. Wide-area networks

4. A/An _____ network is typically a company network that connects multiple company locations into a single network.
   a. Local-area
   b. enterprise
   c. Campus-area
   d. protocol

Section 9.2

5. In a/an _____ networking environment, each user shares its own resources.
   a. enterprise
   b. server-based
   c. peer-to-peer
   d. protocol

6. In a/an _____ networking environment, each user is assigned a username and password by the network administrator.
   a. enterprise
   b. server-based
   c. peer-to-peer
   d. protocol

7. _____ network access will allow others to delete files on your system harddrive.

   a. read-only

   b. server-based

   c. peer-to-peer

   d. full

8. _____ network access will allow others to only see files on your system hard drive.

   a. read-only

   b. server-based

   c. peer-to-peer

   d. full

9. In a peer-to-peer network environment, passwords are always required to access remote devices. True/False

10. In a peer-to-peer network environment, each individual user has his or her own unique username and password for remote device access. True/False

11. Peer-to-peer networks are not designed for large numbers of users. True/False

12. Windows 2000 Server is an example of a _____ network operating system.

    a. noncentralized administration

    b. server-based

    c. peer-to-peer

    d. Stand-alone

13. In a server-based network environment, each individual user has his or her own unique password for remote device access. True/False

14. Server-based networks are not designed for large numbers of users. True/False

Section 9.3

15. _____ are devices used to connect computers to networks.

    a. Drivers

    b. Communications protocols

    c. CDRWs

    d. NICs

16. _____ are pieces of software specific to a hardware device and required for a hardware device to work and interact properly with an operating system.

    a. Drivers

    b. Communications protocols

    c. CDRWs

    d. NICs

17. _____ are pieces of software that format data packets for transmission.

    a. Drivers

    b. Communications Protocols

    c. CDRWs

    d. NICs

18. Ethernet 10Base2 is an example of _____ network topology?

    a. bus

    b. ring

    c. star

    d. stand-alone

19. _____ network topology has all devices attached to a central connection device.

    a. Bus

    b. Ring

    c. Star

    d. Stand-alone

20. _____ network topology has all devices attached to a common backbone or trunk.

    a. Bus

    b. Ring

    c. Star

    d. Stand-alone

21. _____ network topology has the network backbone ends physically connected together.
    a. Bus
    b. Ring
    c. Star
    d. Stand-alone

22. _____ network topology requires termination devices.
    a. Bus
    b. Ring
    c. Star
    d. Stand-alone

Section 9.4

23. The OSI model was developed by the _____.
    a. IEEE
    b. FCC
    c. ISO
    d. TIA

24. The SMTP is an example of a layer-_____ protocol.
    a. 2
    b. 3
    c. 5
    d. 7

25. For communications to occur between common OSI model layers (example: layer 3) on two communicating devices, all OSI model layers are required in sequence. True/False

26. Telnet is an example of a layer-_____ protocol.
    a. 1
    b. 3
    c. 4
    d. 7

27. IPX is an example of a layer-_____ standard.
    a. 2
    b. 3
    c. 5
    d. 6

28. JPEG is an example of a layer-_____ standard.
    a. 1
    b. 2
    c. 6
    d. 7

29. TCP is an example of a layer-_____ protocol.
    a. 1
    b. 3
    c. 4
    d. 7

30. NetBEUI is an example of a layer-_____ standard.
    a. 2
    b. 3
    c. 5
    d. 6

31. The layer-_____ OSI model layer is where Ethernet frames are created.
    a. 1
    b. 2
    c. 6
    d. 7

32. The layer-_____ OSI model layer is the layer where actual bits are put out on the transmission medium.
    a. 1
    b. 2
    c. 6
    d. 7

Discussion Questions

Section 9.1

1. Describe the difference between a local- and campus-area network.

2. Research metropolitan-area networks on the Internet. Can you find any distance limitations for a MAN? If so, what are they? Be sure to reference the site you used in your answer.

3. Is there a campus-area network on your campus? If so, what is the backbone transmission topology (Ethernet, etc.)? What is the backbone bandwidth?

4. Does your campus have Internet access? If so, how much bandwidth does the campus have going off to the Internet? If your answer is in T-carrier units (T-1, etc.) be sure to give the equivalent megabit-per-second rate.

Section 9.2

5. Research peer-to-peer networking with Microsoft Windows products. If possible, set up a peer-to-peer network in your computer lab.

6. Napster and Gnutella applications are considered peer-to-peer applications. How do they differ from an operating system peer-to-peer application (Windows, etc.)? Do you have the option of setting either up for read-only and/or full access? Explain.

7. What kind of servers are you logging into on your campus? Can you log into more than one at a time?

8. Explain why peer-to-peer network passwords are considered "weak."

9. Explain why server-based networks are considered more secure when compared to peer-to-peer network passwords.

10. Use the Internet to research the term *strong password*. What does this term mean? Be sure to list the Web address of your source in your answer.

11. Use the Internet to research general guidelines on managing personal passwords on the Web. List at least five specific guidelines that you, as an individual user on a server-based network, should be using. Be sure to list the Web address of your source in your answer.

Section 9.3

12. Describe two disadvantages of bus topology networks.

13. List three advantages of star topology networks.

14. What kind of network physical topology is being used in your campus labs?

15. Research the history of your campus network—by asking faculty and staff—and create a time line. When was the first network installed? What physical topology was used? Has the physical topology changed? If so, changed to what? and when?

Section 9.4

16. List all seven of the OSI model layers and describe the function of each in your own words.

17. What E-mail application do you use on your campus?

18. Research the RFC process at http://www.faqs.org/rfcs/rfc-index.html What is the most recent RFC on the list? Describe this RFC in your own words.

Connecting LAN Devices

Objectives Upon completion of this chapter, the student should be able to:

- Understand the need for local-area-network standards.
- Discuss the Ethernet 802.3 specification.
- Describe the Ethernet frame layout.
- Discuss Ethernet bus topologies including 10Base2 and 10Base5.
- Define the Ethernet 802.3 5-4-3 rule.
- Define the UTP 568A and 568B cable termination standards.
- Describe the differences between 10BaseT and Fast Ethernet.
- Define the different variations of Fast Ethernet.
- Discuss the differences between Ethernet multiport repeaters and Ethernet switches.
- Describe how an Ethernet switch works.
- Comprehend how Gigabit Ethernet is typically used.
- Define the different variations of Gigabit Ethernet.
- Discuss how Token Ring works.
- Comprehend the concept of a free token in Token Ring networks.
- Define beaconing and how it is used in Token Ring networks.
- Discuss how FDDI works.
- Comprehend the concept of redundancy in FDDI networks.

Key Terms

5-4-3 rule
10Base2
10Base5
10BaseT
100BaseFX
100BaseT4
100BaseTX
568 UTP termination
    standard
568A
568B
802 Standards
    Committee
802.3 Ethernet
802.3z
1000BaseCX
1000BaseLX
1000BaseSX
1000BaseT

beaconing
carrier-sense multiple
    access with collision
    detection
    (CSMA/CD)
collision
collision domain
Fast Ethernet
fiber-distributed-data
    interface (FDDI)
free token
Gigabit Ethernet
jam signal
link segment
media access control
    (MAC) address
medium attachment
    unit (MAU)

medium-dependent
    interface/medium-
    dependent
    interface X
    (MDI/MDIX)
mixing segment
multiport repeater
multistation access unit
    (MAU or MSAU)
repeater
stacking
switch
termination
Token Ring
unshielded-twisted-pair
    (UTP) crossover
    cable
uplinking

Introduction

The two most common data-link-layer protocols used to connect network devices to the network are *Ethernet* and *Token Ring*. This chapter introduces these protocols along with *fiber-distributed-data interface (FDDI)*.

10.1 Ethernet

Ethernet has become the most common method used to build local-area networks due to its relatively low cost, high speed, and reliability. The Ethernet bus network concept was first proposed in 1976 by Dr. Robert Metcalf, who was working for Xerox Corp. at the time. Dr. Metcalf went on to form

3Com Corp. in 1979, a company originally started to develop and sell Ethernet network products. Ethernet has several standards specified by the Institute of Electrical and Electronics Engineers (IEEE: http://www.ieee.org).

Ethernet Standards

The first Ethernet standards were specified by the IEEE in February (month 2) of 1980 and are referred to as the *802 specifications* based on month 2 of year (19)80.

The IEEE maintains the **802 Standards Committee** as the group responsible for developing both local-area-network and metropolitan-area-network information technology standards. Working groups are set up for each major area, and new standards are frequently set. In addition to Ethernet, standards are set for other access methods including security, fiber optics, and wireless communications methods.

Standards are used in telecommunications and other areas to describe a specific process and to provide common ways for devices from different manufacturers to communicate with each other. Standards are critical for mixed-manufacturer communications to work, not just for Ethernet but for all communications protocols and devices. If two individual manufacturers design and build their products to a specific standard, communications between the two products will work properly. Prior to 1980, network communications methods relied on proprietary standards. Mixing devices and connection methods from different manufacturers typically did not work; and, for network communications to work properly, a customer had to purchase communications devices from a single manufacturer. This is illustrated in figure 10-1 and in figure 10-2. Some of the more common IEEE 802 standards are listed in table 10-1. Notice, in addition to Ethernet, that other standards are included.

The IEEE 802.3 Ethernet Specification

The 1980 IEEE **802.3 Ethernet** specification defines **carrier-sense multiple access with collision detection (CSMA/CD)** for the transfer of data from one network device to another. Ethernet devices can be connected in

FIGURE 10-1
Mixed-manufacturer communications failure.

Manufacturer 1 communications solution

Communications failure

Manufacturer 2 communications solution

FIGURE 10-2 Same-manufacturer communications successful.

Manufacturer 1 communications solution

Communications successful

Manufacturer 1 communications solution

TABLE 10-1
Common IEEE 802 standards.

Description	802 specification
LAN/MAN bridging management	802.1
Logical link control	802.2
CSMA/CD access method	802.3
Token passing bus access method	802.4
Token Ring access method	802.5
DQDB access method	802.6
Broadband LAN	802.7
Fiber optics	802.8
Integrated services	802.9
LAN/MAN security	802.10
Wireless LAN	802.11
Demand priority access method	802.12
Cable TV	802.14
Wireless personal-area network (WPAN)	802.15
Broadband wireless access	802.16

either a bus or star configuration and transmit information in the form of data *frames*. The minimum Ethernet 802.3 specification frame size is 512 bits (64 bytes), and the maximum frame size is 12,144 bits (1518 bytes). This means that large pieces of data must be divided into frames for transmission. The Ethernet 802.3 frame configuration is indicated in figure 10-3.

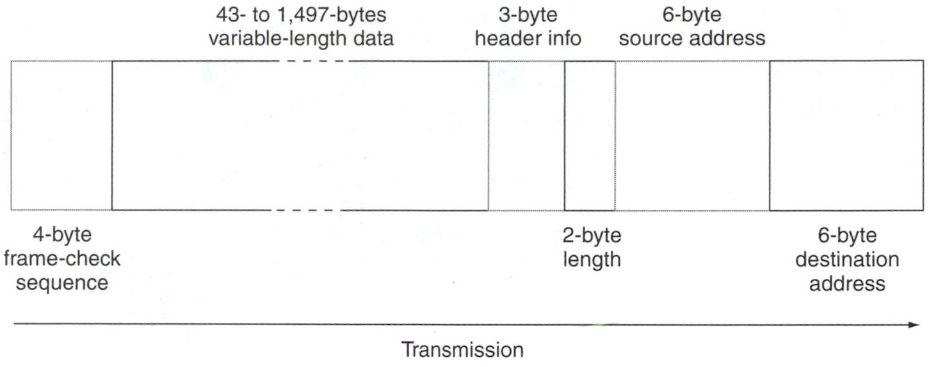

FIGURE 10-3
Ethernet 802.3
frame.

Following are brief definitions for individual frame sections:

Destination address. The address of the network device the frame is going to.

Source address. The address of the network device the frame is coming from.

Header information. Provides frame-handling information.

Variable-length data. The actual data being transmitted from source to destination.

Framecheck sequence (FCS). Also referred to as *cyclic redundancy check (CRC),* the FCS is a calculated number used to check integrity of data transmitted in frame.

Network interface cards (NICs) are hard-coded by the manufacturer with a **media access control (MAC)** address. The *MAC address* is a six-byte unique NIC address used by Ethernet devices to communicate on a LAN. *Carrier-sense multiple access with collision detection (CSMA/CD)* is a network communications method used by Ethernet that allows any device connected on a network to attempt to send a frame at any time. If a connected device has data to transmit, the NIC in the device "listens" to the network and, if the line is not being used, will send a frame. Using this method, the chance always exists that another connected device NIC will transmit a frame at exactly the same time. When this happens, both devices have data to send, both listen and hear nothing, and both send the first frame, as illustrated in figure 10-4.

After a frame is transmitted, a sending Ethernet device NIC will listen for a **collision.** A collision occurs when voltages of two individual pulses from two different attached devices combine, with the resultant voltage double that of the voltage present when one pulse is being transmitted. If a collision is detected on the transmission media, the transmitted frames from colliding devices are discarded and each device waits a random amount of time to

FIGURE 10-4
PC-1 and PC-5
simultaneous frame-
send example.

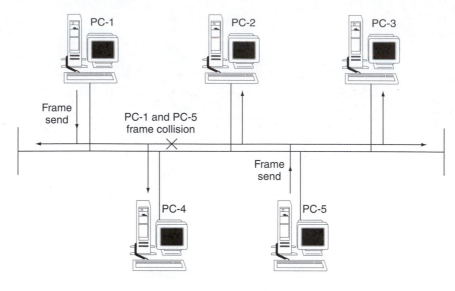

FIGURE 10-5
PC-5 waits and
successfully
retransmits collision
frame while PC-1 still
waits.

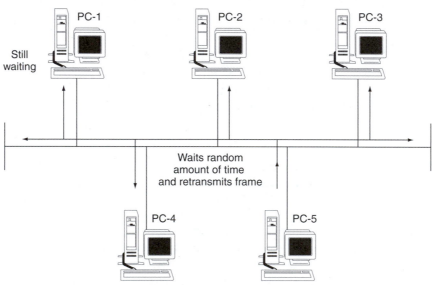

retransmit the same frame again, as indicated in figure 10-5. In the figure, both PC-1 and PC-5 have waited a random amount of time.

Notice in figures 10-4 and 10-5 that an i-beam-like symbol (⊢⊣) is used to represent what is referred to as a **collision domain.** A collision domain is an Ethernet (CSMA/CD) network where collisions will occur when two or more attached devices try to send data at the same time. All devices attached to a single collision domain see all other attached device traffic.

Figure 10-6 indicates that PC-5 has waited less time and retransmits the frame involved in the collision. PC-1 will soon follow after waiting its random time.

FIGURE 10-6
PC-1 successfully retransmits collision frame.

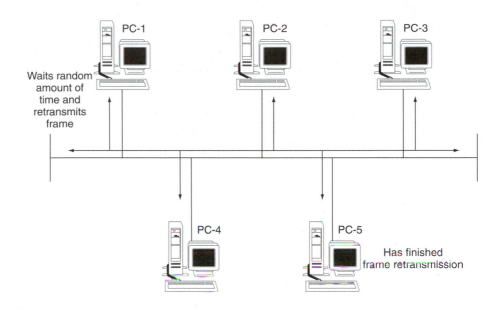

10.2 Ethernet Bus Topology

Coaxial cable was the first transmission medium specified in the original Ethernet standard of 1980. Even though most installations today use twisted-pair media for connections to the desktop, many coaxial local-area-network installations still exist. These installations are more commonly in the **10Base2** configuration and less commonly in the **10Base5** configuration.

10Base2

A common Ethernet bus topology network still used today to connect computers in a local area network configuration is Ethernet *10Base2*. These networks are Ethernet networks that communicate at 10 megabits per second and use thin RG-58 coaxial cable, T-connectors, network interface cards (NICs), and terminating resistors to build a network and attach network devices together. The 10Base2 designation indicates that the network communicates at 10 megabits per second, uses baseband transmission, and has a maximum segment distance limitation of approximately 200 meters (actually 185 meters).

10Base5

In addition to the 10Base2 Ethernet bus network designation is the *10Base5* Ethernet bus network designation. 10Base5 networks are Ethernet networks

FIGURE 10-7
10Base5 AUI device
connection.

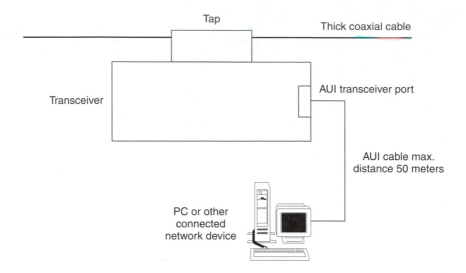

that communicate at 10 megabits per second, use baseband transmission, and have a maximum segment distance limitation of 500 meters. The 10Base5 thick coaxial cable network was the original Ethernet standard developed in 1980.

In these networks, thick RG58, .4-inch-diameter coaxial cable (commonly referred to as *frozen yellow garden hose* because of its size and stiffness) is strung throughout the area to be networked. A 10Base5 transceiver, also referred to as a **medium attachment unit (MAU),** is a device used to physically attach the network device to the thick coaxial cable transmission media. The MAU is clamped onto the thick coaxial cable, and the center conductor inside the cable is vampire tapped by a piercing tap on the transceiver. Once the thick coaxial cable is tapped, an *attachment unit interface (AUI)* is used to attach the computer or other device to the network. The AUI is a 15-pin physical connector interface, and an AUI cable is used to connect the transceiver to an NIC in the computer or other device being attached to the network cable. This is illustrated in figure 10-7. The AUI cable specifications limit AUI cable distance to a maximum of 50 meters with 24-AWG AUI cable used. Less commonly used 28-AWG AUI cable is limited to a maximum length of 16 meters.

Bus Network Termination

10Base2 and 10Base5 network segments require each cable end to be terminated with a 50-ohm resistance. **Termination** is used on Ethernet networks to prevent signals from bouncing off the ends of the coaxial cables and interfering with other transmitted signals. In addition to termination on each end, specifications require one terminating resistance to be earth grounded, as illustrated in figure 10-8. If a terminating resistance is removed or dam-

FIGURE 10-8 Bus network with proper termination on each end.

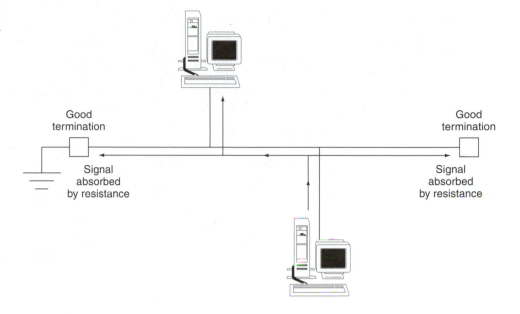

aged, transmitted signals will not be absorbed and will bounce back on the line, interfering with other transmitted signals, as illustrated in figure 10-9.

Bus Network Extension: The Simplified Ethernet 802.3, 5-4-3 Rule

Both 10Base2 and 10Base5 networks can be extended or lengthened using devices called **repeaters.** Repeaters are devices that take any incoming electrical signal on the repeater input line, amplify it, and send it back out on the output line. The Ethernet IEEE 802.3, **5-4-3 rule** is the rule that defines how an Ethernet 802.3 network can have up to five segments connected in series. These series segments are connected with four repeaters and no more than three of the segments can be **mixing segments.**

A *mixing segment* is an Ethernet segment that has greater than two medium-dependent interfaces attached to it. A terminated coaxial cable segment is defined as a mixing segment because additional Ethernet devices (PCs, etc.) can be (but may not be) attached to these segments. **Link segments** are Ethernet segments that cannot have more than two medium-dependent interfaces attached. Most repeaters have built-in medium attachment units (MAUs), and link cables are run directly from repeater to repeater. In figure 10-10, the 10Base2 network is considered fully extended if each of the segments is a maximum length of 185 meters, creating a single collision domain of a maximum length of 925 meters.

FIGURE 10-9 Bus network with missing termination on one end.

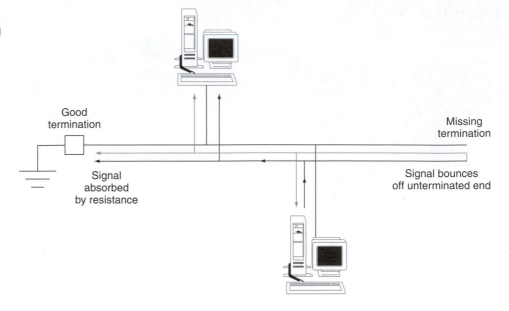

FIGURE 10-10 Ethernet IEEE 802.3, 5-4-3 rule 10Base2 example.

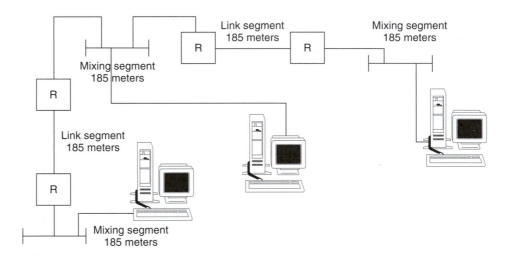

Ethernet System Delay

The five-segment maximum rule defines the maximum length for a single Ethernet 802.3 collision domain. Collision domains can be a mix of coaxial, fiber, and twisted-pair cable with each transmission medium having an associated per meter bit time delay. The Ethernet standard requires a transmitting device to listen for a collision after sending a frame for 50 microseconds. If the physical length of the network is extended beyond the standard five-segment specification the extended length—along with associated delay

FIGURE 10-11
Ethernet IEEE 802.3 extended collision domain data collision.

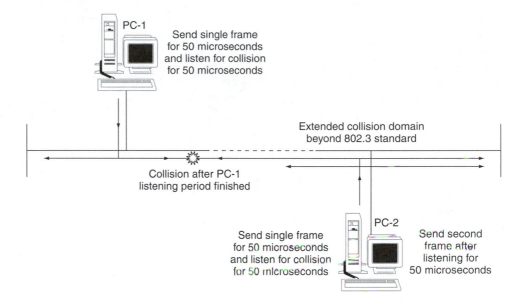

PC-1

Send single frame for 50 microseconds and listen for collision for 50 microseconds

Extended collision domain beyond 802.3 standard

Collision after PC-1 listening period finished

PC-2

Send single frame for 50 microseconds and listen for collision for 50 microseconds

Send second frame after listening for 50 microseconds

times of all attached devices (repeaters, cable types, etc.)—may result in data collision that takes longer than the 50 microseconds to be detected by the transmitting device. This is illustrated in figure 10-11.

10Base2 and 10Base5 Network Disadvantages

Large extended-length bus networks are difficult to maintain. 10Base2 coaxial cable local-area networks were popular for a period of time in the late 1980s and early 1990s, but most new LANs or LANs that have been recently restructured use twisted-pair technology in a star configuration. Making a single 10Base2 PC connection requires the coaxial cable to be cut, British Naval connector or Bayonet Neil-Concelman (BNC) ends installed on each cut, and cable connectivity completed through the T-connector. For these connections to be reliable, high-quality cable, connectors, and terminations must be used at each connection point. An open or shorted cable termination or T-connector will shut down the entire network and can be caused by a number of physical conditions including poor termination, loose connectors, low-quality connectors, or moisture.

When a physical connection causes a network to go down, finding the connection that is bad is typically very difficult since all machines connected cannot communicate, as illustrated in figure 10-12. Typically, technicians end up going from machine to machine jiggling connectors in an attempt to find the one causing the outage. In addition, a disgruntled employee or cleaning person can knowingly or unknowingly shut down a bus network by intentionally disconnecting a connector, bumping a cable attached to a connector, or getting a connector wet. Sometimes barrel connectors are used to

FIGURE 10-12
10Base2 poor
T-connector
connection on a
single-network
device.

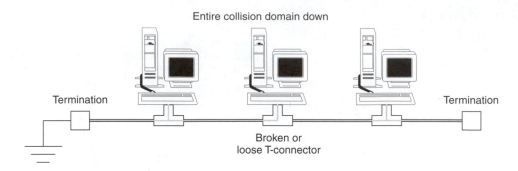

FIGURE 10-13
10Base2 segment
with bad
termination.

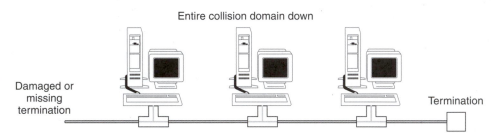

extend the length of a cable and are commonly left lying on the floor. Entire networks have been shut down for extended periods of time due to a connector getting wet from custodial mopping.

Easily removed terminating resistors are another weak point. Without termination on each end, the network will fail, as indicated in figure 10-13.

Bus networks are troublesome and require higher maintenance when compared with other more commonly used topologies. For this reason, most new network installs do not use a bus physical topology to connect directly to Ethernet devices like computers and printers. In modern LAN design, coaxial cable is rarely used exclusively anymore. Some use still occurs in cases in which distances are long and there is no other economical alternative. The classic example is shown in figure 10-14. The sample business here has two buildings separated by a distance that is too long for more commonly used twisted-pair physical topology. In this type of situation, coaxial cable in a bus configuration still may be used to link the networks in each building. However, even these applications are being replaced with more modern and reliable physical topology connection methods like wireless technology.

FIGURE 10-14 Bus configuration linked LANs.

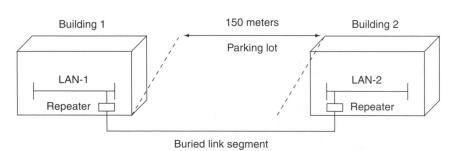

10.3 Twisted-Pair, Data-Grade Cabling

In chapter 4, we learned that many new homes and businesses are having data-grade four-pair unshielded-twisted-pair (UTP) wiring installed for both voice and network access. We also learned the Electronics Industry Association/Telecommunications Industry Association (EIA/TIA) has developed the 568 standard with category 3, 4, 5, and 6 cable allowing the transmission of both voice and data. The 568 standard not only indicates the cable specifications discussed in chapter 4; it also indicates how the cables must be terminated with connectors.

UTP Termination: The 568 Standard

In 1985, the **568 UTP termination standard** was developed by the TIA/EIA in America to provide a wiring standard all manufacturers and installers could work with. Prior to the 568 standard, AT&T had developed a wiring pattern that would work with both networks and two-wire telephones. AT&T named this wiring specification 258A. In 1991, the EIA/TIA standard was released and the AT&T 258A standard was given the name *EIA/TIA 568* by EIA/TIA. In 1994, the EIA/TIA renamed the old 258A standard as *EIA/TIA 568B* and a revised standard was defined as *EIA/TIA 568A*. Both of these standards use eight-wire RJ45 (Register Jack 45) modular connectors similar to the four-wire RJ11 modular connectors used on most modern telephones.

EIA/TIA 568A

The EIA/TIA **568A** pattern is the current standard used for new cable installations in most of the world. The wiring pattern is shown in figure 10-15. The EIA/TIA 568A standard pairs wires as shown in table 10-2.

FIGURE 10-15
EIA/TIA 568A wiring pattern.

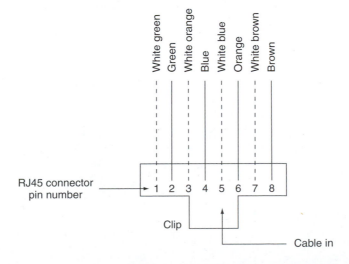

TABLE 10-2
Wire pairs per the
EIA/TIA 568A
standard.

Pair number	Colors	Pins
1	Blue and white blue	4 and 5
2	Orange and white orange	3 and 6
3	White green and green	1 and 2
4	White brown and brown	7 and 8

EIA/TIA 568B

The EIA/TIA **568B** pattern has been replaced by the 568A pattern and is typically only used now to terminate cables that already have a 568B termination. The 568B wiring pattern is shown in figure 10-16. The EIA/TIA 568B standard pairs wires identically to the 568A standard.

Mixing 568A and 568B Standards

Mixing standards on a single cable does not work. RJ45 connectors must be attached to each end of UTP cable, and each end must be terminated using the same 568A or 568B standard. Individual cables, each made to the 568A or 568B standard, can be mixed on a network.

FIGURE 10-16
EIA/TIA 568B wiring
pattern.

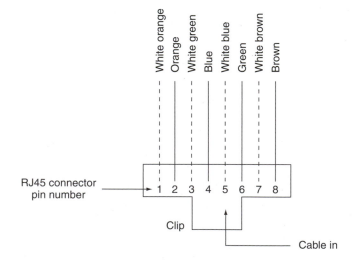

10.4 10BaseT and Fast Ethernet

10BaseT (10-megabits per second baseband over twisted pair) and **Fast Ethernet** (100-megabits-per-second baseband over twisted pair) meet the 802.3 Ethernet specification defined as carrier-sense multiple access with collision detection (CSMA/CD) for the transfer of data from one network device to another with devices attached in a star configuration.

10BaseT

10BaseT Ethernet was the first developed IEEE 802.3 standard for Ethernet over twisted-pair copper wire and is designed to run at 10 megabits per second over category 3, 4, or 5 UTP cable at distances up to 100 meters per device. All connected Ethernet devices are attached to a device called an Ethernet **multiport repeater,** which is also more commonly referred to as an Ethernet *concentrator* or *hub.* See figure 10-17. The 10BaseT specification specifies a 2.5-meter cable length minimum and a maximum 1024 attached network devices per collision domain.

Fast Ethernet

Fast Ethernet is a 100-megabits-per-second Ethernet standard that includes three major versions: **100BaseTX, 100BaseT4,** and **100BaseFX.** Each of these versions is an extension of the original 10BaseX standard. Versions are designed to run over two- and four-pair UTP copper cable and optical fiber. Each is briefly described in the following sections.

FIGURE 10-17
Common 10BaseT network with full extension between two devices.

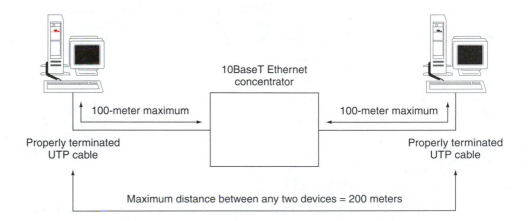

10BaseT Ethernet concentrator

100-meter maximum   100-meter maximum

Properly terminated UTP cable   Properly terminated UTP cable

Maximum distance between any two devices = 200 meters

FIGURE 10-18
100BaseTX sample
network.

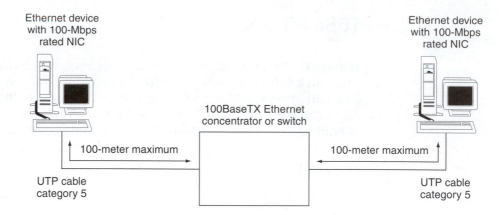

100BaseTX

100BaseTX Ethernet systems are networks designed to run at 100 megabits per second over category 5 UTP cable at distances up to 100 meters per device. Upgrading from a 10BaseT system on category 5 UTP is simply a matter of swapping the Ethernet device NICs and concentrator/switch from the slower 10-megabits-per-second standard to the 100-megabits-per-second 100BaseTX standard. A sample 100BaseTX network is illustrated in figure 10-18. Most Ethernet NICs purchased today support both 10BaseT and 100BaseTX transmission rates. Because the upgrade from 10 megabits per second to 100 megabits per second is so simple and the required devices are inexpensive, 100BaseTX LANs have become extremely popular.

100BaseT4

100BaseT4 Ethernet systems are networks designed to run over four wire pairs at 100 megabits per second over category 3 UTP cable at distances up to 100 meters per device. The 10BaseT4 system was designed so early network cable installs using category 3 cable could use existing cable runs and achieve the higher, Fast Ethernet bandwidth.

Upgrading from a 10BaseT system on category 3 UTP also involves swapping the Ethernet device NICs and concentrator from the slower 10-megabits-per-second standard to the 100-megabits-per-second 100BaseT4 standard, as indicated in figure 10-19. 100BaseT4 NICs and concentrators are typically more expensive than the more popular 100BaseTX devices.

100BASEFX

100BaseFX Ethernet networks use multimode fiber cable to transmit data at 100 megabits per second at distances up to 2 kilometers, as shown in

FIGURE 10-19
100BaseT4 sample network.

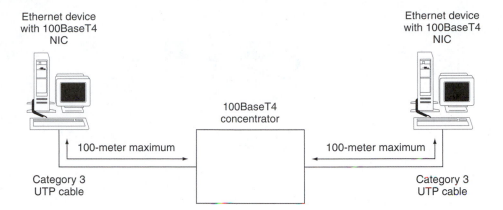

FIGURE 10-20
100BaseFX sample network.

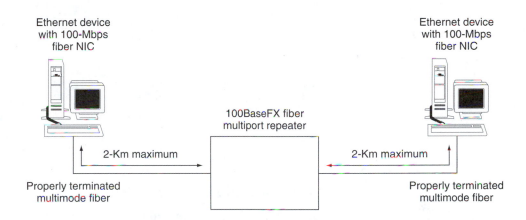

figure 10-20. Fiber termination is relatively expensive when compared to UTP copper wire termination, and the fiber NICs and connection devices are expensive when compared to their UTP counterparts. For this reason, 100BaseFX networks are not as popular as 100BaseTX networks.

10.5 Ethernet Multiport Repeaters versus Switches

Most of our discussion to this point has involved *multiport Ethernet repeaters,* which are also commonly called *concentrators* or *hubs.* **Switches** were briefly mentioned but not discussed. Here, we take a brief look at the major differences between the two.

FIGURE 10-21
Multiport repeater.

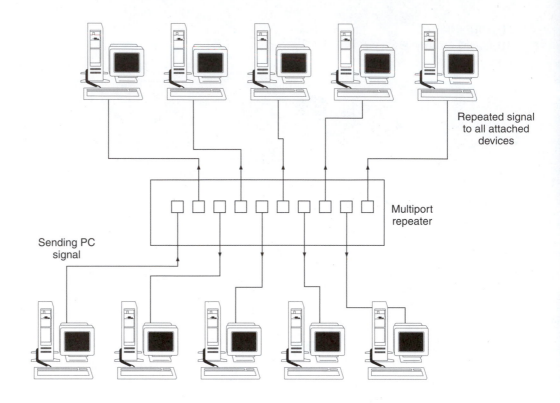

Multiport Repeaters

Ethernet *multiport repeaters* are network devices used to centralize network connections. At the repeater, a signal received on one port is repeated to all attached ports, as illustrated in figure 10-21. A true multiport repeater also detects collisions by listening to transmissions on each port. If two attached devices send data at the same time, the repeater detects the multiple transmissions and sends a **jam signal** to all attached device ports on the repeater. An Ethernet jam signal is a generated signal from one device to all other devices attached to the network that a collision has occurred. The jam signal causes all devices wanting to transmit to restart the CSMA/CD process. This is illustrated in figure 10-22.

Multiport Repeater Stacking

Multiport repeaters may be **stacked** or daisy-chain-connected together to increase the number of Ethernet devices in a single collision domain. The method used to make multiport-repeater-to-multiport-repeater connections

FIGURE 10-22
Multiport repeater
jam signal.

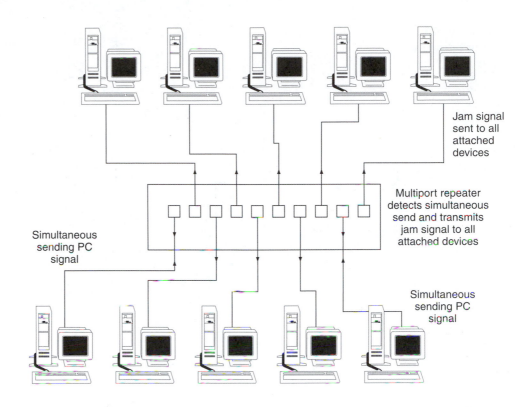

Jam signal
sent to all
attached
devices

Multiport repeater
detects simultaneous
send and transmits
jam signal to all
attached devices

Simultaneous
sending PC
signal

Simultaneous
sending PC
signal

is commonly referred to as **uplinking.** Uplinking can be done using UTP copper, fiber, coaxial cable or special interconnection ports on different manufacturers' devices. A common way to uplink two multiport repeaters uses UTP category 5 568A terminated cables to cross over the multiport repeaters. This can be done in one of two ways. An **unshielded-twisted-pair (UTP) crossover cable** can be constructed or a **medium-dependent interface/ medium-dependent interface X (MDI/MDIX)** pushbutton switch can be used.

Unshielded Twisted-Pair Crossover Cables

An *unshielded-twisted-pair (UTP) crossover cable* is a specially constructed cable that allows Ethernet multiport repeaters to be connected together into a single collision domain. They are commonly constructed using category 5 cable and RJ45 connectors, as outlined in figure 10-23.

Crossover cables can be used to uplink multiport repeaters and can also be used to connect two Ethernet devices without using a hub, as indicated in figure 10-24.

FIGURE 10-23
Ethernet crossover
cable.

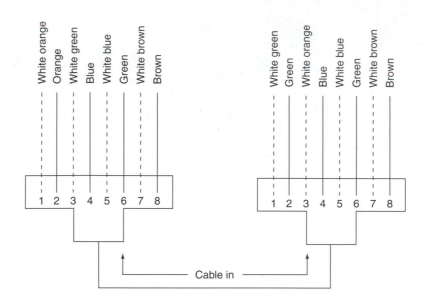

FIGURE 10-24
Ethernet crossover
cable applications.

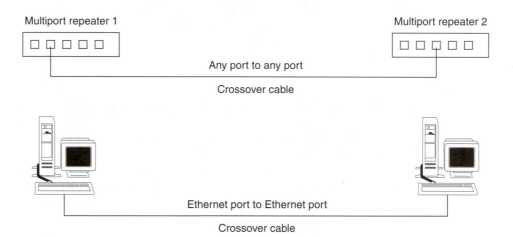

Medium-Dependent Interface/Medium-Dependent Interface X

Most Ethernet multiport repeaters are now manufactured with a *medium-dependent interface/medium dependent interface X (MDI/MDIX)* pushbutton switch that can be used to uplink multiport repeaters using standard UTP network cabling. This switch, when in the MDIX position, connects the MDI/MDIX port in a crossover configuration. The crossing over is done

FIGURE 10-25
Ethernet MDI/MDIX
uplinked multiport
repeaters.

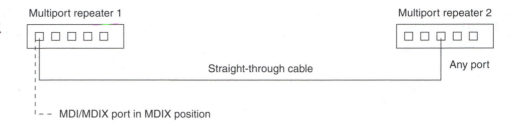

inside the repeater. To make the uplink connection, a standard 568A or 568B cable is used with one end connected into a standard port on one multiport repeater and the other end connected into the MDI/MDIX port on the other multiport repeater, as indicated in figure 10-25. This saves the cable installer from having to make custom cables and having to mark cables as *straight through* or *crossover*.

Different manufacturers specify different maximum port numbers in a stack based on signal propagation delays for each multiport repeater. Remember, in a single Ethernet 802.3 collision domain, each signal must potentially travel the entire collision domain and back in a 50 microsecond period of time. When designing uplinked multiport repeater networks, it is important that the repeaters be matched to prevent missed collision detections.

Ethernet Switches

Ethernet switches resemble multiport repeaters but have the ability to control traffic. *Switches* are essentially enhanced Ethernet network bridges that have been developed to reduce the performance bottlenecks associated with multiport repeaters. Switches are considered intelligent data link layer (layer 2) devices that learn the hardware addresses of the Ethernet device attached to each port as traffic from that device moves from the sending device to the network via the switch. These addresses are stored in a *routing table* (not to be confused with a *router* that also builds and stores a routing table). Figure 10-26 illustrates how a switch learns as data flows through a network. In the figure, PC-1 has the hardware or media access control (MAC) address of 00-03-6D-19-DD-7C and is attached to port 1 of the Ethernet switch. When the switch is first powered up, the routing table is empty. If PC-1 needs to send a file to PC-5, PC-1 will build the first frame and send it out on the network. The switch gets the frame, looks at the sending address, sees the sending address is not in the routing table, and adds the address to the table. The routing table is updated as illustrated in table 10-3. In this example, the switch does not know the addresses of any other attached devices and will

FIGURE 10-26
Ethernet switched
network.

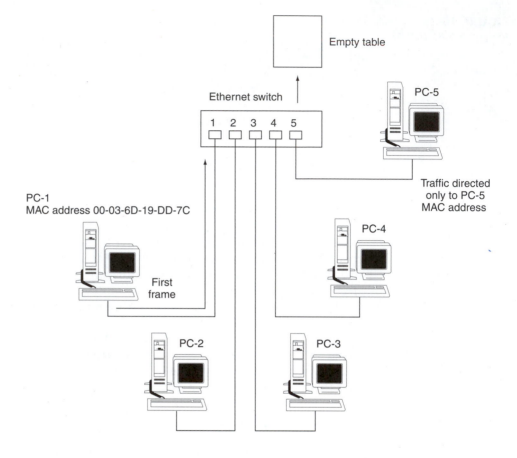

TABLE 10-3 Switch
example showing
single MAC address.

Switch port number	MAC address
1	00-03-6D-19-DD-7C

broadcast the first frame from PC-1 to all ports on the switch without broad-
casting the frame back to PC-1, as shown in figure 10-27.

As each device attached to the switch moves traffic through the switch, the
switch learns the addresses attached to each port and the table will fill with each
device's MAC address. A five-port switch example is shown in table 10-4.
[The MAC addresses are made up by the author.] Once the routing table is
populated, traffic is directed, or forwarded, by the switch to the proper port
based on the frame destination address. This is illustrated in figure 10-28.

FIGURE 10-27
Ethernet switched
network broadcast.

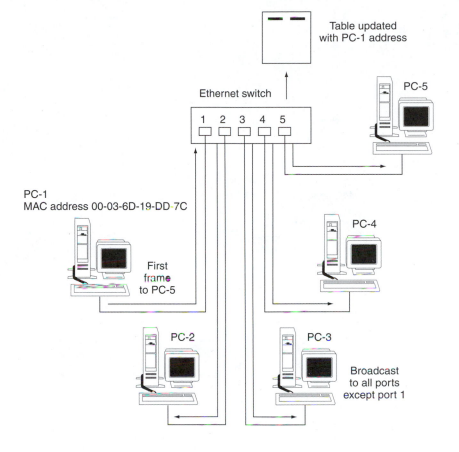

PC-1
MAC address 00-03-6D-19-DD-7C

Table updated
with PC-1 address

Ethernet switch

PC-5

First
frame
to PC-5

PC-4

PC-2

PC-3

Broadcast
to all ports
except port 1

TABLE 10-4
Example of 5-port
switch showing MAC
addresses.

Switch port number	MAC address
1	00-03-6D-19-DD-7C
2	00-03-6D-20-37-A7
3	00-03-6D-77-DE-43
4	00-03-6D-12-34-56
5	00-03-6D-E4-D4-5A

FIGURE 10-28
Ethernet switched
traffic from PC-1
to PC-5.

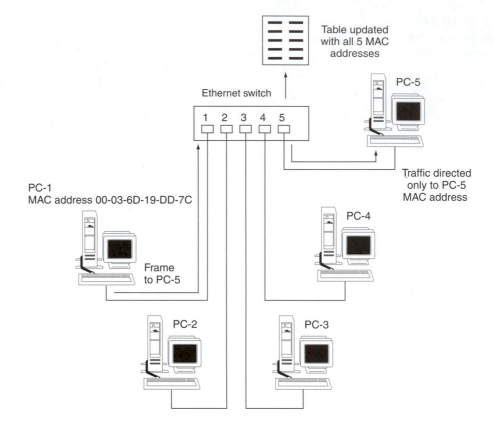

Recall that, in a multiport repeater environment, each device looks at each frame and determines whether the frame is addressed to it. In this switched example, there is no need for PC-1, PC-2, PC-3, or PC-4 to look at traffic directed only to PC-5, allowing these other machines to use processor/ system time to perform other operations resulting in a much more efficient network. Once a switch routing table is populated, the switch actually creates a separate collision domain for every communicating device pair.

In addition to directing traffic to specific devices, switches will not pass invalidly addressed frames that could consume network bandwidth and device processor/system time. Invalidly addressed frames are dropped by the switch, as illustrated in figure 10-29.

Switches essentially treat each port as a separate collision domain, which significantly reduces the number of collisions on a busy network. Most UTP switches sold and installed now are autosensing and have the ability to determine whether the attached device is running at 10 or 100 megabits per second. The individual switch port then adjusts to the proper bandwidth.

FIGURE 10-29
Invalid destination
address dropped by
switch.

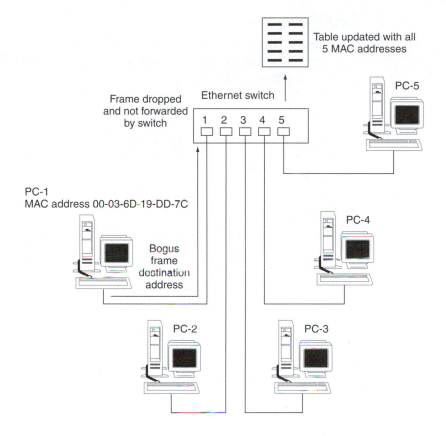

10.6 Gigabit Ethernet

Gigabit Ethernet, also known as the IEEE **802.3z** Ethernet standard, is a standard that originally included three transmission media specifications, two for fiber and one for shielded-twisted-pair copper. Recently, a Gigabit Ethernet specification has been added for UTP category 5 cable. Each specification briefly described in the following sections uses the 802.3 frame format, and each will support a single repeater per collision domain.

1000BaseSX

The **1000BaseSX** Gigabit Ethernet specification operates at 1 gigabit per second and uses short light wavelengths over single-mode fiber cable at distances from 2 to 550 meters.

1000BaseLX

The **1000BaseLX** Gigabit Ethernet specification operates at 1 gigabit per second and uses longer light wavelengths over multimode fiber cable at distances from 2 to 5000 meters.

1000BaseCX

The **1000BaseCX** Gigabit Ethernet specification operates at 1 gigabit per second and uses electrical signals over shielded copper wire at distances up to 25 meters.

1000BaseT

The IEEE has established the 802.3ab Task Force, and the group has specified a **1000BaseT** Gigabit Ethernet standard. 1000BaseT Gigabit Ethernet operates at 1 gigabit per second and allows Gigabit Ethernet to run over all four wire pairs of category 5 unshielded-twisted-pair (UTP) cabling at distances up to 100 meters.

How Gigabit Ethernet Is Used

Gigabit-per-second NICs and centralized gigabit switches, whether fiber or category 5 UTP, are still considered expensive and overkill to run to every device on a LAN. In a LAN environment, typically, a gigabit-per-second connection is made to the servers, the highest traffic volume devices on a LAN, as indicated in figure 10-30. In the figure, the 1-gigabit-per-second fiber connection to the server is shared by all category 5 UTP-connected devices. In this example, ten 10/100-megabit-per-second Ethernet devices are connected to the central switch. Each UTP-connected device can operate at a maximum bandwidth of 100 megabits per second. If 10 devices on the LAN that can transfer at 100 megabits per second are attached to a server that can transfer at 1 gigabit per second, each UTP device could potentially be accessing the server at the 100-megabits-per-second rate at the same time. This is an ideal situation and, in reality, throughput to each PC simultaneously at 100 megabits per second is not likely to happen. This unlikely situation is due not to the network itself but to the individual Ethernet devices, which are, in this case, the 100-megabits-per-second connected PCs and the 1-gigabit-per-second connected server. Individual device central processor unit (CPU) speed and capability, memory (L-1 cache, L-2 cache, and RAM), bus width, and operating system and storage disk speed must be considered when calculating network throughput; and most devices cannot keep up with the 100-megabits-per-second rate. Since in this scenario the fiber-connected server is expected to process and move data at 10 times the rate of the individual UTP-

FIGURE 10-30
Typical gigabit-per-second LAN application.

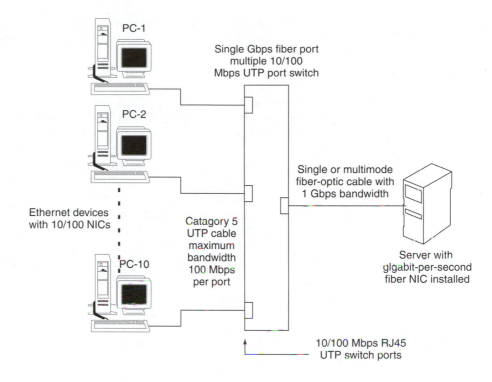

connected devices, the server will be configured with a faster processor, additional memory, a high throughput bus, more robust operating system, and high-speed disk drives.

10-Gigabits-per-Second Ethernet

The latest IEEE 802.3 task force to form has started development of a 10-gigabits-per-second standard referred to as *802.3ae*. This task force is developing specifications for 10-gigabits-per-second Ethernet. Development information can be found at: http://grouper.ieee.org/groups/802/3/ae/index.html

10.7 Token Ring

Token Ring, an IBM (http://www.ibm.com) product, is the most common collapsed-ring topology in use today and is specified with the IEEE 802.5 standard. It uses either shielded-twisted-pair copper wire or UTP data cable. Shielded-twisted-pair (STP) Token Ring configurations commonly use

IBM-type data connector (IDC) hermaphroditic connectors. These connectors do not have gender, and two identical IDC connectors can be connected directly to each other.

More common versions of Token Ring use category 5 UTP cable with RJ45 connectors; there are also Token Ring versions that run over optical fiber. All versions of Token Ring use a collapsed-ring hub called a **multistation access unit (MAU or MSAU).** The Token Ring MAU should not be confused with the Ethernet medium attachment unit, which is also sometimes referred to as an MAU. Most Token Ring network installations are running today at the 16-megabits-per-second standard. Older versions of Token Ring run at 4 megabits per second, and IBM has released a high-speed version that runs at 100 megabits per second. When using UTP, Token Ring will support up to 72 devices per MAU. If using STP, Token Ring will support 260 devices per MAU. Token Ring is considered expensive when compared with the cost of installing a 100-megabit-per-second Ethernet network; for this reason, most new network installs are Ethernet and not Token Ring.

Token Ring Topology

All network devices must be connected directly to the ring formed within the MAU, and a single Ethernet-like crossover cable cannot be used to interconnect two Token Ring MAUs. When two rings are to be joined, each ring must be broken and connections must be made to connect each end of each broken ring. To do this, MAUs have interconnect ports, labeled *ring in* and *ring out,* for extending the size of a ring, as indicated in figure 10-31.

Notice that the extension requires two cables and is typically done using category 5 UTP cable with RJ45 connectors terminated using the EIA/TIA 568A standard. By extending each individual ring this way, the network

FIGURE 10-31
Token Ring
extension.

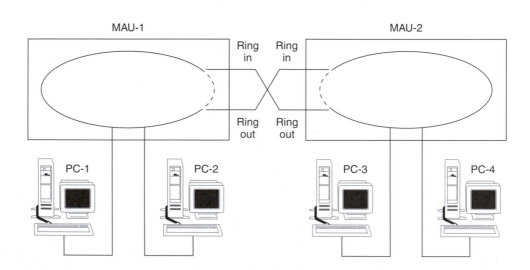

topology now resembles a single large ring with each device attached to that single large ring, as illustrated in figure 10-32.

Rings can be interconnected to form larger networks between departments, floors of a building, or buildings in a campus environment. In a campus configuration, a horizontal ring is run from building to building; vertical rings are run the height of each building; and then horizontal rings are run on individual floors, as illustrated in figure 10-33. At each interconnection point, a MAU exists and ring in/ring out connections are made to connect the rings. A larger campus network like this one can be a mix of transmission media. For example, the campus horizontal ring, or campus backbone, may be optical fiber; the building vertical rings may be STP copper cable; while the floor horizontal rings may use category 5 UTP copper cable.

FIGURE 10-32
Extended large ring.

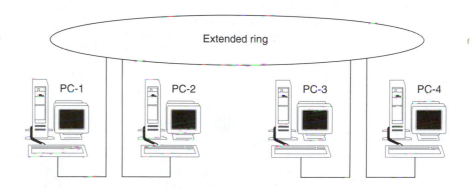

FIGURE 10-33
Extended campus
Token Ring topology.

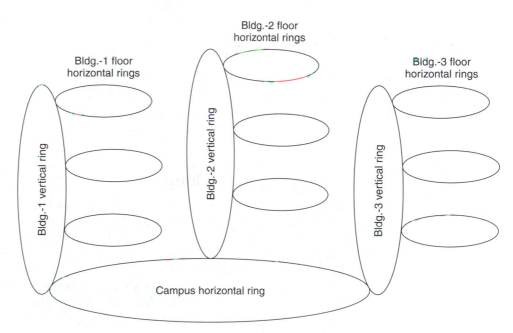

FIGURE 10-34
Token Ring MAU
with attached
powered-down
device.

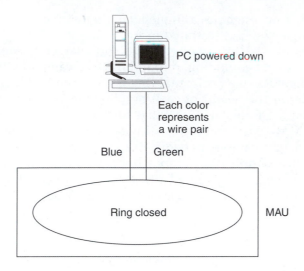

FIGURE 10-35
Token Ring MAU
with attached
powered-up device.

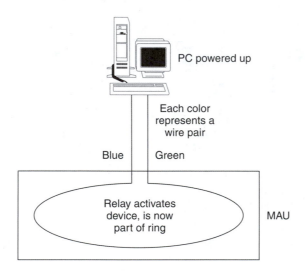

Connecting to the Ring

Token Ring network devices must have a Token Ring NIC installed and properly configured to use the network. The proper cable then must be run from the device to the MAU. The cable uses two pairs of wires to attach to the MAU; one pair is used for transmitting and the other for receiving signals, as illustrated in figure 10-34.

When a device is powered up, a signal voltage is sent to the MAU and the MAU activates a mechanical relay that inserts the device into the ring, as shown in figure 10-35.

FIGURE 10-36
Token Ring free
token flow.

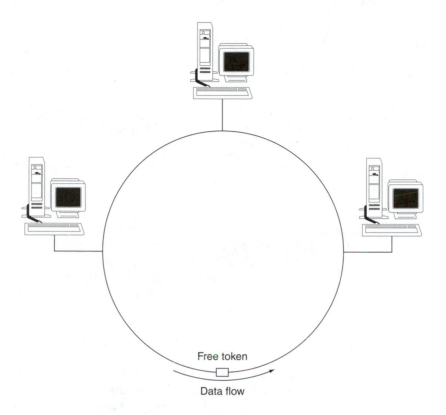

Free token

Data flow

A Token Ring MAU with several attached devices can be quite noisy if all devices are powered up at once and the mechanical relays for each device connection are activated.

Token Passing

Now that we understand how the physical rings are formed, we can look at the way information is transmitted in a Token Ring network. The Token Ring transmission method is significantly different from the CSMA/CD method used by Ethernet networks. Much like Ethernet, only one attached device on a Token Ring network can transmit information at a time; but this is where the similarity stops.

Token Ring networks move a data frame, called a **free token,** around the ring from attached device to attached device, as shown in figure 10-36. This free token allows devices to transmit information on the network. The free token configuration is illustrated in figure 10-37.

If an attached device does not have information to transmit, it passes the free token on to the next device in the ring. If a device has information that needs to be sent, that device picks up the free token. As owner of the free

FIGURE 10-37
Token Ring free
token frame.

Start byte	Access control byte	End byte

FIGURE 10-38
Token Ring busy
token frame
with data.

Start byte	Access control byte	Frame control byte	6-byte destination address	6-byte source address	Variable-length data	4-byte frame-check sequence (FCS)	End byte

token, the device that picks it up is now the only device that can transmit on the ring. The device that picks up the free token changes the token frame format to a busy token format and appends the bits it wishes to send to the token frame. A token frame with data is illustrated in figure 10-38.

A transmitting device can only hold the free token for a set period of time before it has to release the free token to the rest of the network. If a sending station has not finished transmitting when it has to release the free token, it must wait for the free token to circle back before it can continue sending.

As a data frame moves around the ring, it is picked up by each device on the ring and each device analyzes the frame for address information. All frames are checked for errors, regenerated, and passed back out on the ring by all Token Ring devices attached to the ring. If a frame is found with errors, an *error-detection bit* is set in the frame and the frame is put back out on the ring. Other devices will see the set error-detection bit and know the frame is bad. When a good frame gets to the correct destination address device, the frame is saved in memory on the device and two bits are changed on the frame. An *address recognition indicator (ARI) bit* is set to indicate the frame got to the proper destination address, and a *frame-copied indicator (FCI) bit* is set to indicate that the frame was successfully stored in memory on the destination device.

The frame is then sent back out on the ring with these bits set, ultimately returning to the sending device. The sending device analyzes the frame on return and will retransmit if the error-detection, ARI, or FRI bits have not been set properly by the destination device. If the entire block of data has been successfully sent or sending time has elapsed, the free token is regenerated and sent back out on the ring, as illustrated in figure 10-39.

Ring Beaconing

In the event of a connection failure caused by things like device NIC failure or a cable break, ring networks will use **beaconing.** *Beaconing* is an attempt by a device on a ring network to report a problem and reconfigure the net-

FIGURE 10-39
Token Ring successful
transmission data
frame cycle.

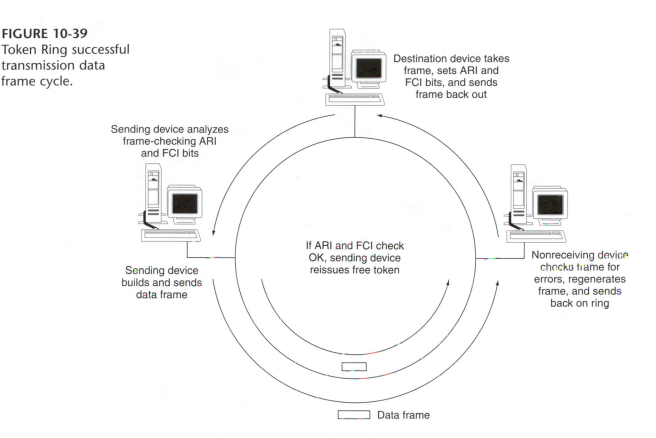

Destination device takes
frame, sets ARI and
FCI bits, and sends
frame back out

Sending device analyzes
frame-checking ARI
and FCI bits

Nonreceiving device
checks frame for
errors, regenerates
frame, and sends
back on ring

Sending device
builds and sends
data frame

If ARI and FCI check
OK, sending device
reissues free token

Data frame

work. Any network device on a ring has the potential to discover a fault or failure in the network. When a device discovers a failure, a beacon frame is sent out on the ring by the first device that detects the failure, as indicated in figure 10-40. In the figure, a network fault is detected by PC-1 and PC-1 sends a beacon frame out on the network. Through a process called *neighbor notification,* each device on the ring knows the address of the device that is immediately upstream. This address is referred to as the *nearest available upstream neighbor (NAUN).* In the figure, the MAU will attempt to reconfigure and bypass the fault using the relay on the connected port. In the event of a destroyed PC-4 connection, PC-4 will be eliminated from the ring by the MAU. PC-1 will continue to send beacon frames until it sees a beacon frame from itself. When PC-1 does receive a beacon frame from itself, it will assume the network failure is repaired and will stop sending beacon frames and generate a new free token so network communications can continue.

The Future of Token Ring

Token Ring networks are not commonly installed now due primarily (whether actually true or not) to equipment expense and the general concept

FIGURE 10-40
Token Ring
beaconing.

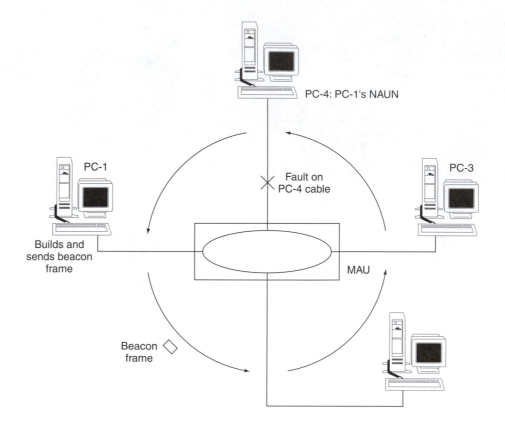

that only 4- and 16-megabits-per-second versions exist. IBM has released a high-speed Token Ring 100-megabit-per-second product, but switched 100BaseT is currently the most common network topology installed. Token Ring has some advantages, with the most common being statistical probability. Network designers can more easily predict overall network performance of Token Ring networks than Ethernet networks because the Token Ring networks follow a much more ordered set of rules. Most LANs do not require the guaranteed level of performance that Token Ring offers, and Fast Ethernet has become the preferred method.

10.8 Fiber-Distributed-Data Interface

Fiber-distributed-data interface (FDDI) networks, developed in the early 1990s, use a dual-counter rotational-ring configuration to connect network devices at 100 megabits per second in a CAN or MAN environment. Two fiber lines are run from device to device, and network devices can be

FIGURE 10-41
Fiber-distributed-data interface configuration.

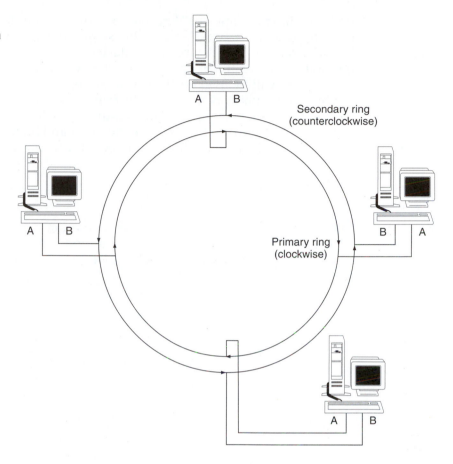

dual-attached with two connections per network device, as illustrated in figure 10-41. Each device has two ports labeled *A* and *B*. The primary ring is attached to port A and the secondary to port B. In addition to dual port devices, M-port devices exist that only have a single port that connects only to the primary ring.

An FDDI network runs over optical fiber, is primarily used as a network backbone, and transports data at 100 megabits per second. Up to 500 network devices can be attached to an FDDI ring, and ring lengths can be up to 100 kilometers. In addition to FDDI, a copper wire unshielded-twisted-pair (UTP) version is called *copper-distributed-data interface (CDDI)*. The FDDI networks were commonly used for large campus backbones, and the rings are typically buried in separate locations so that, in the event of physical ring damage (e.g., backhoe fade), the network will continue to operate.

The communications method is similar to Token Ring, with a free token traveling around the ring; when a device has something to send, the free token is held and the information is sent in the form of an FDDI frame.

Included in the frame is a destination address; each device on the network picks up the frame, looks at the address, and determines whether the frame is addressed to that device. Each device also retransmits the frame in a method similar to Token Ring; in the event of a primary ring failure, the secondary ring will be used until the primary is repaired.

Most new network backbone installs are no longer using FDDI. Other technologies, such as asynchronous transfer mode (ATM), Gigabit Ethernet, or even Fast Ethernet, offer higher bandwidth and better overall performance when mixing data, voice, and multimedia traffic on the same network.

Summary

1. The two most common local-area-network protocols are *Ethernet* and *Token Ring.*

2. The IEEE maintains the *Ethernet 802 Standards Committee* that is responsible for setting network connectivity standards.

3. The IEEE 802.3 specification defines *carrier-sense multiple access with collision detection (CSMA/CD).*

4. Ethernet information is transmitted across the network in the form of *frames.*

5. Ethernet frames are constructed at the *data-link layer (layer 2)* of the OSI model.

6. Ethernet devices use *media access control (MAC)* addresses to communicate across a local-area network.

7. *10Base2* Ethernet networks operate at 10 megabits per second at distances up to 185 meters.

8. *10Base5* Ethernet networks operate at 10 megabits per second at distances up to 500 meters.

9. Ethernet bus networks require that *terminating resistors* be added to the network cable ends.

10. *Repeaters* are physical layer (layer 1) devices that take an incoming signal, amplify it, and send it back out on the network. They are used to extend the physical length of a network.

11. The Ethernet *5-4-3 rule* defines how an Ethernet network can be extended.

12. Ethernet bus networks that use coaxial cable are commonly being replaced with *UTP cable networks* in a star configuration.

13. The *568 UTP termination standard* has been developed by the TIA/EIA to provide a standard that all manufacturers and installers can work with.

14. *10BaseT* Ethernet networks operate at 10 megabits per second at distances up to 100 meters using UTP cable.

15. Fast Ethernet is a 100-megabits-per-second Ethernet standard.

16. *100BaseTX* Ethernet networks operate at 100 megabits per second at distances up to 100 meters per device using category 5 UTP cable.

17. *100BaseT4* Ethernet networks operate at 100 megabits per second at distances up to 100 meters per device using category 3 and higher UTP cable.

18. *100BaseFX* Ethernet networks operate at 100 megabits per second at distances up to 2 kilometers per device using multimode fiber cable.

19. Ethernet *multiport repeaters* are network repeaters used to centralize network connections.

20. A *crossover cable* or *MDI/MDIX* pushbutton switch can be used to stack Ethernet multiport repeaters.

21. *Ethernet switches* are intelligent data link layer (layer 2) devices that learn the MAC addresses of all devices directly attached to them.

22. *Gigabit Ethernet* is also known as the IEEE *802.3z* specification and operates at 1 gigabit per second.

23. *1000BaseSX* Ethernet networks operate at 1 gigabit per second at distances up to 550 meters per device using short light wavelengths over multimode fiber-optic cable.

24. *1000BaseLX* Ethernet networks operate at 1 gigabit per second at distances up to 5000 meters per device using long light wavelengths over multimode fiber-optic cable.

25. *1000BaseCX* Ethernet networks operate at 1 gigabit per second at distances up to 25 meters per device using shielded copper cable.

26. *1000BaseT* Ethernet networks operate at 1 gigabit per second at distances up to 100 meters per device using category 5 UTP cable.

27. *Token Ring* is the most common collapsed-ring, local-area-network technology used today.

28. Token Ring networks use a collapsed-ring hub called a *multistation access unit (MAU* or *MSAU)*.

29. Token Ring extension is done using the *ring in* and *ring out* ports on a Token Ring MAU.

30. Token Ring frames are constructed at the *data-link layer (layer 2)* of the OSI model.

31. Token Ring networks use a data frame called a *free token* to allow attached devices to communicate on the ring.

32. Token Ring networks use *ring beaconing* to report a network configuration problem and attempt to reconfigure a network around a problem.

33. Token Ring networks are being *replaced* with faster and less-expensive Ethernet local-area networks.

34. *Fiber-distributed-data Interface (FDDI)* networks are used to connect network devices in a CAN or MAN environment and run over optical fiber at 100 megabits per second at distances up to 100 kilometers.

35. The FDDI networks use a *dual-counterrotational-ring configuration* with each ring similar to that of Token Ring.

36. Each network device on an FDDI network has *two network connections* with one of the connections commonly used as a backup.

37. Most new CAN and MAN installs *no longer use FDDI;* other technologies like ATM, Gigabit Ethernet, and Fast Ethernet are typically being used.

Review Questions

Section 10.1

1. _____ is the most common method currently used to build LANs.
   a. Token Ring
   b. FDDI
   c. Ethernet
   d. ATM

2. The IEEE _____ Standards Committee is responsible for developing all Ethernet standards.
   a. 802
   b. 100-Mbps
   c. ITU
   d. 232

3. _____ are used in the telecommunications industry to define how devices work together in a mixed-manufacturer environment.

   a. Cables

   b. Standards

   c. Terminators

   d. Links

4. Two proprietary standard devices from two different manufacturers typically have no problems communicating with each other. True/False

5. _____ addresses are used by network devices to communicate on a LAN.

   a. ITU

   b. CCITT

   c. CSMA/CD

   d. MAC

6. _____ is a network communications method used by Ethernet that allows any device connected to a network to attempt to send a frame at any time.

   a. 802

   b. Token passing

   c. CSMA/CD

   d. Frame relay

7. An Ethernet network pulse _____ results in the doubling of pulse transmission voltage.

   a. collision

   b. avoidance

   c. retransmission

   d. wait

Section 10.2

8. _____ Ethernet networks run at 10 megabits per second at distances up to 185 meters.

   a. 10BaseT

   b. 10Base2

   c. 100BaseT

   d. 10Base5

9. _____ Ethernet networks run at 10 megabits per second at distances up to 500 meters.

   a. 10BaseT

   b. 10Base2

   c. 100BaseT

   d. 10Base5

10. Ethernet bus networks perform better with termination on both ends but will still work as long as only one end of the bus is properly terminated. True/False

11. Ethernet bus networks can be physically extended using the _____ rule.

    a. gigabit-per-second

    b. 5-4-3

    c. 3-2-1

    d. fast

12. Ethernet _____ take an incoming signal, amplify it, and send it back out on the network.

    a. doublers

    b. terminators

    c. repeaters

    d. segmenters

13. An Ethernet _____ segment has greater than two medium-dependent devices attached to it.

    a. mixing

    b. link

    c. bus

    d. series

14. An Ethernet _____ segment cannot have greater than two medium-dependent devices attached to it.

    a. mixing

    b. link

    c. bus

    d. series

15. A transmitting 10-megabits-per-second Ethernet device will listen for a collision after sending a frame for _____ microseconds.

    a. 10

    b. 20

    c. 25

    d. 50

Section 10.3

16. The current UTP EIA/TIA termination standard is _____ .

    a. 568A

    b. 568B

    c. 568C

    d. 568D

17. A UTP category 5 cable terminated on one end as 568A and the other as 568B will not function properly on a local-area-network. True/False

Section 10.4

18. The _____ version of Ethernet was the first Ethernet standard that ran over twisted-pair copper wire.

    a. 10Base2

    b. 10Base5

    c. 10BaseT

    d. 100BaseT4

19. _____ Ethernet runs at 100 megabits per second only on category 5 cable.

    a. 10BaseT

    b. 100BaseFX

    c. 100BaseTX

    d. 100BaseT4

20. _____ Ethernet runs at 100 megabits per second only on category 3, 4, or 5 cable using four wire pairs.

    a. 10BaseT

    b. 100BaseFX

    c. 100BaseTX

    d. 100BaseT4

21. _____ Ethernet runs at 100 megabits per second over multimode fiber-optic cable.

    a. 10BaseT

    b. 100BaseFX

    c. 100BaseTX

    d. 100BaseT4

Section 10.5

22. Ethernet _____ are network devices that receive signals and always repeat these signals to all attached devices.

    a. MAUs

    b. AUIs

    c. switches

    d. multiport repeaters

23. Some Ethernet multiport repeaters will generate and transmit a _____ signal to indicate a data collision on the network.

    a. jam

    b. clog

    c. collide

    d. negative

24. Connecting Ethernet multiport repeaters together to increase the number of Ethernet devices in a single collision domain is commonly referred to as _____.

    a. jamming
    b. uplinking
    c. moving
    d. combining

25. _____ can be used to connect multiport repeaters together. Choose two.

    a. MDI/MDIX pushbutton switches and UTP cables
    b. 50-ohm terminators
    c. RJ11 jacks
    d. UTP crossover cables

26. Ethernet switches build a table of _____ to keep track of all attached devices.

    a. IP addresses
    b. MAC addresses
    c. computer names
    d. computer model numbers

27. An Ethernet multiport repeater typically provides better performance than an Ethernet switch in a LAN environment. True/False

Section 10.6

28. Gigabit Ethernet is also referred to as the IEEE _____ standard.

    a. 802.5
    b. 802.11
    c. 802.3z
    d. 802.3ae

29. The _____ Gigabit Ethernet specification uses single-mode fiber at distances up to 550 meters.

    a. 1000BaseT
    b. 1000BaseLX
    c. 1000BaseCX
    d. 1000BaseSX

30. The _____ Gigabit Ethernet specification uses multimode fiber at distances up to 5000 meters.

    a. 1000BaseT
    b. 1000BaseLX
    c. 1000BaseCX
    d. 1000BaseSX

31. The _____ Gigabit Ethernet specification uses shielded copper wire at distances up to 25 meters.

    a. 1000BaseT
    b. 1000BaseLX
    c. 1000BaseCX
    d. 1000BaseSX

32. The _____ Gigabit Ethernet specification uses UTP category 5 cable at distances up to 100 meters.

    a. 1000BaseT
    b. 1000BaseLX
    c. 1000BaseCX
    d. 1000BaseSX

33. 10 Gigabit Ethernet is currently under development in the IEEE _____ standard.

    a. 802.5
    b. 802.11
    c. 802.3z
    d. 802.3ae

Section 10.7

34. Token Ring networks use a collapsed-ring hub called a/an _____.

    a. MAU
    b. AUI
    c. MSU
    d. UMA

35. Most Token Ring network installations today run at _____ megabits per second.

   a. 4

   b. 10

   c. 16

   d. 24

36. Token Ring networks use a _____ to indicate that a device can send information on the network.

   a. contention scheme

   b. free token

   c. blank frame

   d. mechanical relay

37. Token Ring networks use _____ to indicate a connection or device failure.

   a. ring collisions

   b. ring contentions

   c. ring in and ring out

   d. ring beaconing

Section 10.8

38. FDDI network devices commonly have _____ connection/connections per networking device.

   a. 1

   b. 2

   c. 2 pair

   d. 4 pair

39. FDDI networks commonly run at _____.

   a. 4 Mbps

   b. 10 Mbps

   c. 100 Mbps

   d. 1 Gbps

Discussion Questions

Section 10.1

1. Research the IEEE 802 Standards Committee on the Internet. What is the current fastest standard under development? Describe this standard.

2. If your computer is connected to the Internet, determine, depending on the operating system you are using, how to find the MAC address of the NIC in the machine. What procedure do you use and what is the address?

3. Define CSMA/CD. If too many devices are on a CSMA/CD-based network, is it theoretically possible for the network to saturate to the point where no devices can ever communicate? Explain.

4. Define a collision in an Ethernet network in terms of voltage.

Section 10.2

5. Two buildings are 300 meters away from each other with a conduit pipe connecting the two. Budgets are tight, and you want to connect the two buildings together. You have access to surplus 10Base2 and 10Base5 equipment. How could you inexpensively connect the buildings?

6. Why are coaxial-cable-based Ethernet networks no longer commonly installed? Give at least three reasons, and explain each one.

7. Why is termination required in a bus-based network?

8. Describe the Ethernet 5-4-3 rule. What would be the length of a fully extended 10Base5 network using the 5-4-3 rule?

9. How long will a sending Ethernet device wait to detect a collision? Describe how extending the physical length of a network beyond the standard limitations can cause problems for devices to detect collisions.

Section 10.3

10. Describe the difference between 568A and 568B UTP termination.

11. Explain why 568A and 568B termination cannot be used on the same cable but can be mixed when multiple cables are used.

Section 10.4

12. Can 10BaseT run over category 3 UTP? Explain.

13. Can 100BaseTX run over category 3 UTP? Explain.

14. The term *autosensing* is used when referring to modern Ethernet NICs. What does this term mean?

15. Explain why fiber is typically not run to the desktop computer NIC.

Section 10.5

16. How does a multiport repeater function? Would you consider it to be a true repeater based on the earlier repeater definition? Explain.

17. Explain how an Ethernet jam signal works. Does your lab have any Ethernet multiport repeaters? If so, are do they generate jam signals?

18. Construct a crossover cable in the lab, and try connecting two multiport repeaters together. Does it work?

19. Describe the basic differences between an Ethernet multiport repeater and a switch.

Section 10.6

20. Describe how Gigabit Ethernet is commonly used. Is there any Gigabit Ethernet on your campus? If so, to what is it connected?

21. On the IEEE Web site, research the 802.3ae standard and give an update of its status.

Section 10.7

22. Describe how a collapsed ring is formed in a Token Ring MAU.

23. How are rings extended using Token Ring? Can a single cable be used to extend a ring? Explain.

24. Describe the process of Token Ring token passing. What is meant by the terms *free token* and *busy token?* Explain.

25. Describe the process of ring beaconing used by Token Ring to indicate a fault in the ring.

26. If too many devices are on a Token Ring network, is it theoretically possible for the network to saturate to the point where no devices can ever communicate? Explain.

Section 10.8

27. How similar to Token Ring is FDDI? Explain.

28. Explain why FDDI is commonly not used as a backbone technology any more. What has typically replaced it?

Telecommunications Filters

Filter theory is an important thing to understand when studying the existing public switched telephone network (PSTN). Also referred to as the plain old telephone system (POTS), the PSTN is the existing telephone network we can all use to communicate with other users throughout the world. Early in the development of the voice network, engineers realized the importance of different frequency bandwidths. **Frequency** is defined as the number of cycles an alternating signal completes in 1 second. The standard unit of frequency (f) is the **hertz (Hz);** one hertz is the equivalent of one cycle per second, as illustrated in figure A-1.

1 Hz = 1 cycle per second

The time it takes for one cycle to occur is the period T. Period and frequency are reciprocals:

$$f = \frac{1}{T} \quad \text{and} \quad T = \frac{1}{f}$$

Filters are electronic devices that allow for the selection and corresponding suppression of specific frequency ranges. *Bandwidth* is defined as the width (in hertz) of a band of frequencies in a communications system that can be used to transmit information. As the need to transmit higher volumes of information from one location to another has grown, our needs for bandwidth have grown.

FIGURE A-1
Alternating signal.

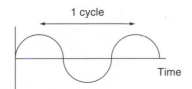

Low Pass Filters

Low-pass filters allow signals with lower frequencies to pass from input to output while rejecting higher frequencies. Some basic filter definitions can be identified while referring to figure A-2. These definitions can be applied to any filter type.

> *Passband.* The low-pass-filter **passband** is defined as the range of low frequencies passed by a low-pass filter within a specified limit that is determined by the *cutoff frequency.*
>
> *Reject region.* The low-pass-filter **reject region** is defined as the range of higher frequencies not passed by a low-pass filter; the reject limit is also determined by the *cutoff frequency.*
>
> *Cutoff frequency.* The low-pass-filter **cutoff frequency** (f_c) is the frequency at which a low-pass filter's output voltage is 70.7 percent of its maximum value. Cutoff frequency is also called the *break frequency.*

Resistor Capacitor Circuit Low-Pass Filter

A series resistor capacitor (RC) circuit low-pass filter is a series RC circuit with the output taken across the **capacitor.** This circuit is also referred to as an *RC lag network* and is illustrated in figure A-3.

FIGURE A-2 Low-pass-filter frequency response.

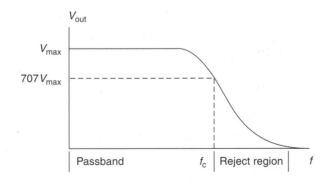

FIGURE A-3 RC circuit low-pass filter.

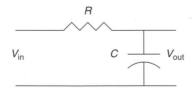

FIGURE A-4
RC circuit low-pass
filter with 20-volt
DC source.

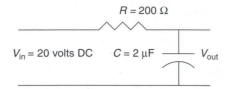

DC Voltage Source

The frequency response of a series RC circuit is illustrated in figure A-4 starting with a direct current (DC) source. A DC source is a source that has a frequency of 0 hertz. In the figure, V_{in} is a 20-volt DC source. Since it is a DC source, the frequency (f) of the input is 0 ($f = 0$) and the output voltage (V_{out}) is equal to the value of the input voltage:

$$V_{out} = V_C = 20 \text{ volts DC}$$

Why? Recall **capacitive reactance (X_C),** which is the equivalent resistance of a capacitive device at a specific frequency. It is calculated as follows:

$$X_C = \frac{1}{2\pi f C}$$

Equation A-1

If the frequency is 0 for the DC source ($f = 0$) and we substitute into the equation, then:

$$X_C = \frac{1}{2\pi(0)C} = \frac{1}{0} = \infty$$

The capacitor, under DC conditions ($f = 0$), resembles an open circuit of infinite resistance and is said to *block* constant DC. As a result, no current exists in the circuit and no voltage drop occurs across the resistor. The circuit passes all of the input voltage to the output:

$$V_{out} = V_C = 20 \text{ volts DC}$$

AC Voltage Source with 500-Hertz Frequency

Now, using the same resistor and capacitance values for the circuit in figure A-4, we can switch V_{in} to an alternating current (AC) source with a value of 10 volts$_{rms}$ (root mean square) and a frequency of 1 kilohertz, as illustrated in figure A-5.

$$V_{in} = 20 \text{ volts}_{rms} \quad \text{and} \quad \text{Input frequency} = 500 \text{ Hz}$$

$$R = 200 \text{ }\Omega$$

$$C = 2 \text{ }\mu\text{F}$$

FIGURE A-5
RC circuit low-pass
filter with 500-hertz
AC source.

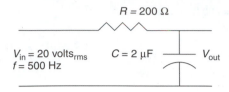

Since we are dealing with a capacitor, which is a reactive device whose value changes as frequency changes, we need to find V_{out} using a complex voltage divider equation:

$$V_{out} = V_C = \left[\frac{X_C}{\sqrt{R^2 + X_C^2}} \right] V_{in}$$

Equation A-2

In this example, we know:

$V_{in} = 20 \text{ volts}_{rms}$ and $R = 200 \ \Omega$

but we do not know the value of X_C. We can calculate it as follows:

$$X_C = \frac{1}{2\pi f C}$$

If

$f = 500 \text{ Hz}$

and we substitute into the equation, then:

$$X_C = \frac{1}{2\pi(500 \text{ Hz})(.000002F)} = 159 \ \Omega$$

Now we can substitute into the complex voltage divider equation and calculate the output voltage:

$$V_{out} = V_C = \left[\frac{X_C}{\sqrt{R^2 + X_C^2}} \right] V_{in} = \left[\frac{159 \ \Omega}{\sqrt{200 \ \Omega^2 + 159 \ \Omega^2}} \right] 20 \text{ volts}_{rms}$$

$$= 12.44 \text{ volts}_{rms}$$

Here, with an input frequency of 500 hertz, only 12.44 volts or 62.2 percent of the input voltage is actually reaching the output.

FIGURE A-6
RC circuit low-pass filter with 5-kilohertz AC source.

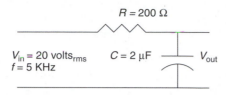

AC Voltage Source with 5-Kilohertz Frequency

We can now take the same circuit in figure A-4 and figure A-5 and adjust the input frequency to a value of 5 kilohertz, as shown in figure A-6:

$$V_{in} = 20 \text{ volts}_{rms} \quad \text{and} \quad \text{Input frequency} = 5 \text{ KHz}$$

$$R = 200 \ \Omega$$

$$C = 2 \ \mu F$$

Once again, to find V_{out}, we need to use a complex voltage divider:

$$V_{out} = V_C = \left[\frac{X_C}{\sqrt{R^2 + X_C^2}} \right] V_{in}$$

Here:

$$V_{in} = 20 \text{ volts}_{rms} \quad \text{and} \quad R = 200 \ \Omega$$

Once again, we do not know the value of X_C but can calculate it using:

$$X_C = \frac{1}{2\pi f C}$$

Using a frequency of 5 kilohertz, we can substitute into this equation:

$$X_C = \frac{1}{2\pi(5 \text{ KHz})(.000002F)} = 15.9 \ \Omega$$

Using the new value of X_C, we can substitute again into the complex voltage divider equation:

$$V_{out} = V_C = \left[\frac{X_C}{\sqrt{R^2 + X_C^2}} \right] V_{in} = \left[\frac{15.9 \ \Omega}{\sqrt{200 \ \Omega^2 + 15.9 \ \Omega^2}} \right] 20 \text{ volts}_{rms}$$

$$= 1.58 \text{ volts}_{rms}$$

FIGURE A-7
RC circuit low-pass filter frequency response curve.

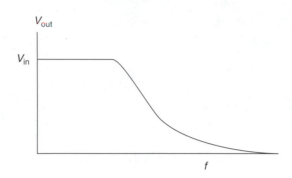

With an input frequency of 5 kilohertz only 7.9 percent of the input voltage is actually reaching the output. If we were to continue to increase the input frequency, X_C would approach 0 and V_{out} would keep decreasing until it equaled approximately 0 volts. This is illustrated in the frequency response curve shown in figure A-7.

A series RC circuit with the output taken across the capacitor is called a low-pass filter. Low frequencies are allowed to pass to the output while the higher frequencies are not.

Resistor Inductor Circuit Low-Pass Filter

A resistor inductor (RL) circuit low-pass filter is a series resistor inductor circuit with the output taken across the resistor, as illustrated in figure A-8. Understanding of this circuit is extremely critical for telecommunications students. The copper wire pair, referred to as the local loop, that most of us have coming into our homes has been designed and tuned as an RL circuit low-pass filter.

FIGURE A-8
RL circuit low-pass filter configuration.

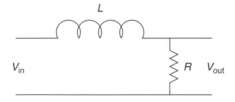

Recall **inductive reactance (X_L)** is the equivalent resistance of an **inductor** at a specific frequency. It is calculated as follows:

$$X_L = 2\,\pi f L$$

Equation A-3

Looking at equation A-3, we can see that, at lower frequencies, X_L will be small and the majority of input will be dropped across R and appear at the output. As the input frequency increases, X_L increases, and more of the input

FIGURE A-9
RL circuit low-pass filter with DC voltage source.

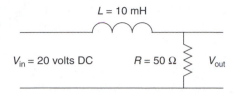

voltage will be dropped across L and less will be dropped across R. Less of the input voltage appears at the output; for this reason, inductors are said to **choke** high frequencies.

DC Voltage Source

The frequency response of a series RL circuit starting with a DC source is illustrated in figure A-9. Recall that a DC source has a frequency of 0 hertz. In the figure, we see that V_{in} is a 20-volt DC source. Since it is a DC source, the frequency (f) of the input is zero ($f = 0$), and we can say that output voltage (V_{out}) is equal to the value of the input voltage.

Why? Recall inductive reactance (X_L):

$$X_L = 2\pi fL$$

If $f = 0$ and we substitute into the equation, then:

$$X_L = 2\pi(0)L = 0$$

The inductor resembles a short circuit under DC conditions, so no voltage drop occurs across the inductor. All of the input voltage is dropped across the resistor and appears at V_{out} and:

$$V_{out} = V_R = V_{in} = 20 \text{ volts DC}$$

AC Voltage Source with 500-Hertz Frequency

Now, using the same resistor and inductor values as in figure A-9, let us switch V_{in} to an AC source with a value of 20 volts$_{rms}$ and a frequency of 500 hertz as shown in figure A-10.

$V_{in} = 20$ volts$_{rms}$ and   Input frequency = 500 Hz

$R = 50 \ \Omega$

$L = 10 \text{ mH}$

FIGURE A-10
RL circuit low-pass filter with 2-kilohertz input.

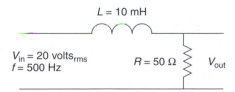

Inductors, just like capacitors, are reactive devices that change with frequency. To find V_{out}, a complex voltage divider must be used:

$$V_{out} = V_C = \left[\frac{R}{\sqrt{R^2 + X_L^2}} \right] V_{in}$$

Equation A-4

We know:

V_{in} = 20 volts$_{rms}$ and $R = 50\ \Omega$

and must calculate the value of X_L using:

$X_L = 2\ \pi fL$

If f = 500 hertz and we substitute into the equation, then:

$X_L = 2\ \pi fL = 2\pi(500\ Hz)(.01\ H) = 31.41\ \Omega$

Knowing the value of X_L, we can substitute into the complex voltage divider equation:

$$V_{out} = V_C = \left[\frac{R}{\sqrt{R^2 + X_L^2}} \right] V_{in} = \left[\frac{50\ \Omega}{\sqrt{50\ \Omega^2 + 31.41\ \Omega^2}} \right] 20\ volts_{rms}$$

$$= 16.93\ volts_{rms}$$

We can see at a frequency of 500 hertz, 16.93 volts, or 84.7 percent of the input voltage, reaches the output.

AC Voltage Source with 2-Kilohertz Frequency

We can now increase the frequency of the input to 2 kilohertz for the circuit in figure A-9 and figure A-10, resulting in the circuit in figure A-11. In the figure:

V_{in} = 20 volts$_{rms}$ and   Input frequency = 2 KHz

$R = 50\ \Omega$

$L = 10\ mH$

FIGURE A-11
RL circuit low-pass filter with 2-kilohertz input.

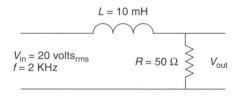

Again, to find V_{out} we need to use a complex voltage divider:

$$V_{out} = V_C = \left[\frac{R}{\sqrt{R^2 + X_L^2}}\right]V_{in}$$

We know:

$V_{in} = 20$ volts$_{rms}$ and $R = 50\ \Omega$

We must calculate X_L

$X_L = 2\ \pi f L$

If $f = 6$ kilohertz and we substitute into the equation, then:

$X_L = 2\ \pi f L = 2\pi(2000\ Hz)(.01\ H) = 125.66\ \Omega$

Notice that X_L is larger at 2 kilohertz than it was at 500 hertz. Now we can substitute into the complex voltage divider equation:

$$V_{out} = V_C = \left[\frac{R}{\sqrt{R^2 + X_L^2}}\right]V_{in} = \left[\frac{50\ \Omega}{\sqrt{50\ \Omega^2 + 125.66\ \Omega^2}}\right]20\ volts_{rms}$$

$$= 7.39\ volts_{rms}$$

Now only 7.39 volts, or 36.9 percent, of the input voltage is actually reaching the output. If we were to continue to increase the input frequency, X_L would approach infinity and V_{out} would keep decreasing until it equaled approximately 0 volts. This is illustrated in the graph in figure A-12.

An RL circuit with the output taken across the resistor is called a low-pass filter. Low frequencies are allowed to pass to the output while the higher frequencies are not.

FIGURE A-12
RL circuit low-pass-filter frequency response curve.

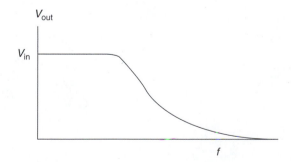

Glossary

5-4-3 rule   Rule that defines how an Ethernet 802.3 network can have up to five segments connected in series.

10Base2   Ethernet networks that communicate at 10 megabits per second and use thin RG-58 coaxial cable, T-connectors, network interface cards (NICs), and terminating resistors to build a network and attach network devices together.

10Base5   Ethernet networks that communicate at 10 megabits per second and use thick RG-8 coaxial cable, MAUs, AUIs, network interface cards (NICs), and terminating resistors to build a network and attach network devices together.

10BaseT   Ethernet networks that run at 10 megabits per second over category 3, 4, or 5 UTP cable at distances up to 100 meters per device.

100BaseFX   Ethernet networks that use multimode fiber cable to transmit data at 100 megabits per second at distances up to 2 kilometers.

100BaseT4   Ethernet networks designed to run over four wire pairs at 100 megabits per second over category 3 UTP cable at distances up to 100 meters per device.

100BaseTX   Ethernet networks designed to run at 100 megabits per second over category 5 UTP cable at distances up to 100 meters per device.

568 UTP termination standard   TIA/EIA local-area-network wiring standard.

568A   Current UTP local-area-network wiring and termination standard that replaced the 568B standard.

568B   Old UTP local-area-network wiring and termination standard that has been replaced by the 568A standard.

802 Standards Committee   IEEE committee that is responsible for setting local area network (LAN) communications standards.

802.3 Ethernet   IEEE specification that defines Ethernet.

802.3z   IEEE Gigabit Ethernet standard.

1000BaseCX   Gigabit Ethernet specification that operates at 1 gigabit per second and uses electrical signals over shielded copper wire at distances up to 25 meters.

1000BaseLX   Gigabit Ethernet specification that operates at 1 gigabit per second and uses longer light wavelengths over multimode fiber cable at distances from 2 to 5000 meters.

1000BaseSX   Gigabit Ethernet specification that operates at 1 gigabit per second and uses short light wavelengths over single-mode fiber cable at distances from 2 to 550 meters.

1000BaseT   Gigabit Ethernet specification that operates at 1 gigabit per second and allows Gigabit Ethernet to run over all four wire pairs of category 5 unshielded-twisted-pair (UTP) cabling at distances up to 100 meters.

acknowledgment (ACK)   A reply back from a receiving data terminal to a sending data terminal indicating that a piece of information was successfully received.

American National Standards Institute (ANSI)   A private, nonprofit organization that administers and coordinates the U.S. voluntary standardization and conformity assessment system.

American Standard Code for Information Interchange (ASCII)   Common format used to transmit data between computers; all characters are represented using a seven-bit binary number.

American wire gauge (AWG)   A U.S. numbering standard for wire size with "gauge" indicating wire diameter.

amplitude-shift keying (ASK)   A technique used to convert digital signals to analog signals by modulating signal amplitude.

analog modem   A modem designed to work on the public switched telephone network.

application layer   Layer 1 of the OSI model with two major functions: it provides a common interface to program applications, and it determines if the required resources are available for communications to occur.

ARPANet   The U.S. defense network started in the 1960s that has now become the Internet.

asymmetric digital subscriber line (ADSL)   A technique used by telephone companies for high-bandwidth broadband data communications over standard telephone wires.

asynchronous transfer mode (ATM)   Connection-oriented digital switching technique that uses 53 byte cells to transmit information.

ATM cell header   Section of ATM cell that identifies the cell and allows the cell to be routed through an ATM network on the correct virtual channel.

automatic repeat request (ARQ))   A reply back from a receiving terminal to the sender indicating that a piece of information that was sent needs to be retransmitted; also referred to as an *automatic retransmission request.*

automatic retransmission request (ARQ)   A reply back from a receiving terminal to the sender indicating that a piece of information that was sent needs to be retransmitted; also referred to as an *automatic repeat request.*

average frequency   The average value of frequency in a variable-frequency system.

band-pass filter   Electronic filter designed to only allow a range of frequencies to pass.

band-stop filter   Electronic filter designed to allow all frequencies to pass except a specified range.

bandwidth   The width (in hertz) of a band of frequencies in a communications system.

baseband   A communications system that transmits information on a single channel.

basic-rate interface (BRI)   The integrated services digital network (ISDN) lower-bandwidth service that offers two digital 64-kilobits-per-second B-channels for voice and data transmission and one digital 16-kilobits-per-second D-channel for signal transmission.

baud   A transmission bandwidth term. One baud is the equivalent of one signal change or cycle per second allowing the transmission of a single bit of information per cycle. Today, transmission techniques allow multiple bits to be transmitted per cycle.

B-channel   The integrated services digital network (ISDN) channel used to transmit information.

beaconing   An attempt by a device on a ring network to report a problem and reconfigure the network.

binder   Wiring grouping indicator. For example, in a 25-pair binder, five groups of five wire pairs are combined .

bipolar with 8-zero substitution (B8ZS)   T-1 circuit method used to maintain transmitting and receiving device synchronization.

bit robbing   The single bit between the sixth and twelfth frames a T-1 carrier master frame used for in-band signaling.

bits per second (bps)   A transmission bandwidth term indicating the number of bits that can be transmitted by a transmission system in one second of time.

blocking   Term used when all telephone switch connections are in use and a call is not allowed to complete.

bridged tap   An unterminated wire pair that has been tapped into to provide telephone service to a customer.

broadband   A communications system that transmits information on multiple channels simultaneously.

broadband ISDN (B-ISDN)   ISDN-provided services in a broadband environment.

bus topology   Network physical configuration that attaches all devices (PCs, printers, etc.) to a single cable that is commonly referred to as the network *backbone* or *trunk*.

busy hour   Used by telecommunications traffic engineers, is defined as the hour with the most call traffic in the specified time period. During this hour, some calls may be blocked if switch capacity is exceeded.

caller ID   Telephone service that displays a caller's name and telephone number on receiving telephone equipment.

calling party number (CPN)   Caller ID information placed .5 second after the first ring of a telephone call by the SS7 network.

campus-area networks (CANs)   Computer networks that are used to interconnect LANs in buildings that are in close proximity to each other.

capacitive reactance (X_C)   Reactive resistance of a capacitor; calculated using the formula: $X_C = \dfrac{1}{2\pi f C}$

capacitor   Electronic device that stores energy and then returns it to an electronic system. Device consists of two conductors separated by an insulator referred to as a *dielectric*. Units of measure are farads (F).

carrierless amplitude phase (CAP)   Digital subscriber line modulation technique that uses all of the available bandwidth as a single channel and then optimizes the bit rate to the existing line conditions.

carrier-sense multiple access with collision detection (CSMA/CD)   A network communications method used by Ethernet that allows any device connected on a network to attempt to send a frame at any time.

caveat   A legal notice filed in court requesting the postponement of a proceeding until the filing party has had the chance to be heard.

cell   With reference to ATM, the 53-byte fixed-length packet used to transmit information on an ATM network.

central exchange (CE)   A telephone company facility that contains telephone switching equipment. *Central office* has the same meaning.

central office (CO)   A telephone company facility that contains telephone switching equipment. *Central exchange* has the same meaning.

century call second (CCS)   Unit used to monitor and measure telephone network traffic. To calculate the CCS value, the call count is multiplied by the average time for each call; this number is then divided by 100.

choke   Term commonly used for inductance.

client   A connected network device.

coaxial cable   Copper-core cable surrounded by insulation and a conductive shield. In a communications system, the core carries the signal and the shield protects from interference and functions as ground.

CODEC   An acronym for coder/decoder; typically, a single chip on a modem that converts digital signals to analog and analog signals to digital.

collapsed ring   Ring network topology with the ring formed inside a connection hub.

collision   With reference to Ethernet, occurs when two different attached network devices transmit pulses at the same time and the voltages of the two individual pulses are added together.

collision domain   An Ethernet (CSMA/CD) network where collisions will occur when two or more attached devices try to send data at the same time. All devices attached to a single collision domain see all other attached device traffic.

Commit Consultatif International Telegraphique et Telephonique (CCITT)   Also referred to as the International Telephone and Telegraph Consultative Committee, an international standard-setting organization currently known as the ITU-T (International Telecommunications Union telecommunication standardization sector) located in Geneva, Switzerland.

committed information rate (CIR)   With reference to frame relay, the guaranteed throughput bit rate for a frame-relay PVC.

communications protocols   Piece of software that sets communications rules that allow two devices to talk to each other on a network.

community antenna television (CATV)   The cable television system, now commonly referred to as *community access television* or simply *cable TV.*

companding   Used to compress and divide lower-amplitude transmission signals into more voltage levels and provide more signal detail at the lower voltage amplitudes.

competitive local-exchange carrier (CLEC)   A company that competes with a local telephone company providing telephone service (e.g., a cable television company providing dialtone in competition with the local telephone company).

consumer asymmetric digital subscriber line (CADSL)   A technology that allows existing telephone lines to be used for both voice and high-speed data at the same time.

crossbar switch   An electromechanical telephone switching system that uses a matrix of individual switch connections.

crosstalk   Electrical noise caused by electromagnetic fields on one conductor being electromagnetically coupled onto another conductor in close proximity.

custom local area signaling services (CLASS)   Enhanced voice services including caller ID, call return, repeat dialing, priority ringing, select call forwarding, call trace, and call blocking.

cutoff frequency   Frequency at which an electronic filter's output voltage is 70.7 percent of its maximum value. Cutoff frequency is also called the *break frequency,* and the output is said to be down 3 decibels at this frequency.

cyclic redundancy check (CRC)   Method used to detect transmission errors. Technique involves a calculation based on sending device data. Receiving device takes same data and performs same calculation. If calculation results match, then transmission occurred correctly.

data compression   Communications technique that reduces the number of bits that need to be transmitted in a communications session; saves time and money.

data-link layer   Layer 2 of the OSI model that provides reliable error-free delivery of data over the network transmission medium.

dBm   Telecommunications power measurement unit that is referenced to a 1 mW power level.

D-channel   The ISDN channel used for signal transmission.

decibel (dB)   A logarithmic measurement of voltage or power ratios that is used to express input-to-output relationships.

defacto standard   A standard that has not been approved by a standards organization but has been accepted and used by an industry as a standard.

demodulation   The conversion of analog signals to digital signals that a receiving digital device can understand.

dense wavelength division multiplexing (DWDM)   Very high capacity multiplexing technique that uses wavelength, or color, of light to combine signal channels onto a single piece of optical fiber.

dialtone   telephone signal formed by the simultaneous transmission of a 330 hertz tone and a 440-hertz tone.

digital multiplexing   The technique of combining digital signals typically using one of two techniques—time division multiplexing (TDM) and statistical time division multiplexing (STDM).

digital subscriber line (DSL)   Telecommunications service that allows high-bandwidth transmission to homes and businesses over the existing local loop telephone lines. There are several versions of DSL (ADSL, HDSL, etc.) typically, the *xDSL* designation is used when discussing one specific form of the technology.

digital subscriber line access multiplexer (DSLAM)   The device to which a local loop is connected that provides DSL Internet service. The device multiplexes multiple customer data connections onto a high-bandwidth Internet connection.

discrete multitone (DMT)modulation   A digital subscriber line (DSL) signaling method that separates the DSL frequency range into 256 separate frequency channels (or channels) of 4.3125 kilohertz each.

dispersion   The separation of light waves as they travel down a piece of fiber-optic cable.

distribution point   Cable network device used to serve a typical neighborhood and move the television signals off of a trunk cable onto smaller feeder cables.

drop cables   Cable network cables that tap into a feeder cable and enter the customer premises.

DS-0   A 64-kilobits-per-second channel that is the basic building block for the existing digitally multiplexed T-carrier system in the United States and the digital E-carrier system used in Europe.

DS-1   Twenty-four 64-kilobits-per-second digital DS-0 voice channels multiplexed together to form a bit rate of 1.544 megabits per second when combined with added transmission overhead.

DS-2   Four, 1.544-megabits-per-second digital DS-1 signals multiplexed into one signal to form a bit rate of 6.312 megabits per second when combined with added transmission overhead.

DS-3   Seven, 6.312-megabits-per-second digital DS-2 signals multiplexed into one signal to form a bit rate of 44.736 megabits per second when combined with added transmission overhead.

DS-4   Six, 44.736-megabits-per-second digital DS-3 signals multiplexed into one signal to form a bit rate of 274.176 megabits per second when combined with added transmission overhead.

DS-5   Two, 274.176-megabits-per-second digital DS-4 signals multiplexed into one signal to form a bit rate of 560.16 megabits per second when combined with added transmission overhead.

DSL lite   Digital subscriber line (DSL) service that runs at 1.5 megabits per second downstream and up to 512 kilobits per second upstream. Also known as *splitterless DSL* or *universal DSL*.

dual-tone-multifrequency (DTMF)   The most commonly used telephone signaling method for inputting a number in the United States and Europe. DTMF phones typically use a 12-button keypad. When a button is pressed on the keypad, an electric contact is closed and two oscillators generate two tones at specific frequencies.

E-carrier   European digital transmission format that is slightly different than the North American T-carrier system format.

electronic switching system (ESS)   Computerized telephone switching system that uses reed switches to connect and disconnect calls.

Electronics Industry Association/Telecommunications Industry Association (EIA/TIA)   Dual organization that is accredited by the American National Standards Institute (ANSI) to develop voluntary industry standards for a wide variety of telecommunications products.

encoding   The process of taking a piece of an analog signal, quantizing and companding it, and then giving it an 8-bit binary code.

encryption   Process of scrambling data so that only authorized receivers can unscramble the data.

enterprise network   A network that connects multiple LANs that are geographically distributed within an organization.

Erlang   Unit that telephone company traffic engineers use to denote telephone traffic.

Ethernet   The most commonly used local-area-network (LAN) technology in use today. Ethernet is specified by the IEEE 802.3 standard.

exchange   The central location where local loop wire pairs come together and are integrated into the telephone network. Early exchanges contained a manual switchboard and human operators who switched calls. Modern exchanges are computer controlled.

far-end crosstalk (FEXT)   Electromagnetic signal crossover that occurs between two signals transmitted in the same direction on a copper wire pair.

Fast Ethernet   A 100-megabits-per-second Ethernet standard that includes three major versions: 100BaseTX, 100BaseT4, and 100BaseFX.

Feature Group D   Signaling method that allows customers to bypass their preferred long-distance carrier by dialing a *101XXXX* number and using another long-distance carrier.

Federal Communications Commission (FCC)   U.S. organization that sets telecommunications standards.

feeder cables   Cable network cables that are run through residential neighborhoods.

fiber-distributed-data interface (FDDI)   Network method that uses a dual counter rotational fiber ring configuration to connect network devices at 100 megabits per second in a CAN or MAN environment.

filters   Electronic devices that allow for the selection and corresponding suppression of specific frequency ranges.

firewall   A device used to prevent unauthorized access to a specified network.

frame   With reference to T-carriers, a combination of 24 digital sample signals with a single framing bit added. A frame is 193 bits in length. With reference to Ethernet and Token Ring, a data unit created at the data-link layer of the OSI model (e.g., Ethernet frame).

frame relay   A packet-switching networking technology designed to transfer traffic between data networks.

framing   With reference to T-carriers, the process of adding one bit to a 192-bit T-1 carrier bit pattern to create a 193-bit frame.

free token   Token Ring network data frame that allows devices to transmit information on the network.

frequency   The number of cycles an alternating signal completes in 1 second.

frequency division multiplexing (FDM)   Multiplexing technique that uses different frequencies to represent different communications channels with transmission done on copper or microwaves.

frequency multiplexing   Analog multiplexing technique used until the early 1990s by long-distance carriers in the United States.

frequency-shift keying (FSK)   Modulation technique that varies the frequency of a carrier signal.

full access   A network sharing option that allows others attached to the network to read, download, save, and delete files in the shared folder.

G.lite   Digital subscriber line (DSL) service that runs at 1.5 megabits per second downstream and up to 512 kilobits per second upstream; also known as *universal ADSL* or *splitterless DSL*.

geosynchronous earth orbit   A satellite orbit that allows a satellite to follow the earth and stay over the same location on Earth.

grade of service   Used by traffic engineers to indicate the probability that a call will be blocked at a given century call second (CCS) rate.

graded index fiber   Fiber-optical transmission medium with more bandwidth and less loss than step-index fiber but lower bandwidth and higher loss than single-mode fiber; not commonly used in high-bandwidth telecommunications systems.

group   A frequency division multiplexed combination of 12 voice call channels.

hacker   Someone who manipulates computer systems and networks in a destructive way.

headend   Location where all cable network system signals are received for delivery on the local cable network. The headend typically consists of one or more satellite links and a building housing transmission equipment.

hertz (Hz)   Frequency units; one hertz is the equivalent of one cycle per second.

high-bit-rate digital subscriber line (HDSL)   Digital subscriber line service that offers symmetric 1.544 megabits per second over two copper wire pairs and symmetric 2.048 megabits per second over three wire pairs. HDSL local loop distance is restricted to 1.2 kilometers, but phone companies can install signal repeaters to extend this range up to 12 kilometers.

high-bit-rate digital subscriber line 2 (HDSL2)   Digital subscriber line service that offers symmetric 2 megabits per second over a single copper pair. Service uses pulse amplitude modulation (PAM) and became a standard on May 21, 2001.

high-level data link control (HDLC)   A group of communications protocols that function at layer 2 of the OSI model and are responsible for transmitting communications traffic through communications nodes.

high-pass filter   An electronic filter that allows signals with higher frequencies to pass from input to output while rejecting lower frequencies.

Huffman encoding   Data-compression technique that uses a frequency table and categorizes each symbol (or character) in a character string.

hybrid fiber coaxial (HFC)   Combination fiber and coaxial cable system that provides a cable bandwidth of 700 megahertz and allows delivery of up to 110 different television channels to the customer.

impedance   A resistive value for a circuit at a specified electronic frequency.

in-band signaling   Signaling technique that uses the telephone voice connection for network signaling.

index of refraction   Optical material mathematical ratio of c/v where c is the speed of light in a vacuum and v is the speed of light in the specified optical material.

inductor   Electronic device that stores energy and then returns it to an electronic system. Inductors can be discrete devices or the result of wire loops in a system.

inductive reactance (X_L)   Reactive resistance of an inductor, calculated using the formula: $X_L = 2\pi f L$.

inside wire (IW)   Common residential telephone wiring that has four copper wires that combine to form two wire pairs; also referred to as *station wire, POTS wire,* and *JK wire.*

insulator   Nonconducting device used to prevent and protect against unwanted electric current flow.

integrated services digital network (ISDN)   Worldwide standard for an all-digital network that uses existing local loop lines to carry digital signals to ISDN digital switches in a local CO.

interexchange carrier (IXC)   Long-distance telephone service provider selected by the customer.

interLATA call   Considered a long-distance call between two telephone companies' geographical areas that is switched to the customer's selected long-distance carrier point of presence (POP).

International Telecommunications Union (ITU)   International telecommunications standards setting group located in Geneva, Switzerland.

International Telecommunications Union (ITU-T)   The telecommunication standardization sector of the ITU, an international telecommunications standards-setting group located in Geneva, Switzerland. The ITU-T was formerly known as the CCITT.

Internet Engineering Task Force (IETF)   An international organization that controls how communications protocols are set up. (See: http://www.ietf.org)

Internet protocol (IP) address   Unique 32-bit binary address assigned to any network device attached to a network that uses the TCP/IP protocol suite. The most commonly used TCP/IP network is the Internet.

Internet service provider (ISP) Company that provides customer Internet access; connection can be wire, fiber, or wireless.

intraLATA call A non-long-distance call within a single telephone company's geographical area.

ISDN digital subscriber line (IDSL) Digital subscriber line service that uses the same 2B1Q modulation code as ISDN to deliver service without special line conditioning. IDSL is a nonswitched service.

jam signal Ethernet multiport-repeater-generated signal that indicates that a collision has occurred.

jumbogroup A frequency division multiplexed combination of 3600 individual voice channels.

jumbogroup multiplex A frequency division multiplexed combination of 10,800 individual voice channels.

laser An acronym for *light amplification by stimulated emission of radiation.* Laser light is a coherent beam of electromagnetic energy with all output having the same frequency and phase.

laser diode A semiconductor device that produces laser light.

L-carrier Obsolete transmission system developed by AT&T in the 1970s that used coaxial cable to carry multiplexed voice conversations from telephone center to center.

link segments Ethernet segments that cannot have more than two medium-dependent interfaces.

load factor Used by telecommunications traffic engineers, indicates the average number of telephone calls that can be made at the same time during the defined busy hour.

loading coil Inductive device added to the local loop to cancel the effects of capacitive shunting.

local access and transport area (LATA) U.S. term that defines a geographical area covered by one or more local telephone companies.

local loop The connection between a telephone being used in a residence or business and the local-exchange carrier (LEC) central office (CO); also referred to as the *subscriber loop.*

local-area network (LAN) Simple computer network consisting of a collection of computers and peripherals in a small area connected together to share resources.

local-exchange carrier (LEC) The telephone company responsible for the final telephone network connection coming into a home or business.

logical link control (LLC) sublayer The upper of the two OSI model data-link sublayers; identifies the specific communications line protocol.

low-pass filter An electronic filter that allows signals with lower frequencies to pass from input to output while rejecting higher frequencies.

masterframe Grouping of 12 frames in the DS carrier system; also referred to as a *superframe*.

mastergroup A frequency division multiplexed combination of 600 individual voice channels.

media access control (MAC) address The six-byte unique NIC address used by Ethernet devices to communicate on a LAN.

media access control (MAC) sublayer The OSI model lower data-link sublayer responsible for physical connection sharing among computers connected to the network.

medium attachment unit (MAU) Ethernet 10Base5 transceiver used to physically attach the network device to the thick coaxial cable transmission media.

medium-dependent interface/medium-dependent interface X (MDI/MDIX) Pushbutton switch that can be used to uplink Ethernet multiport repeaters using standard UTP network cabling.

metropolitan-area network A network used to connect remote LANs, usually located in the same city, into an enterprise network.

microwave Wireless electromagnetic energy having a frequency higher than 1 gigahertz.

mixing segment An Ethernet segment that has greater than two medium-dependent interfaces attached to it.

modem Device that takes a digital signal from a computer or other digital device and converts or modulates it into an analog voice frequency for transmission on the public switched telephone network (PSTN). Once received, a modem on the other end takes the analog signal and demodulates, or converts, it back to digital.

modulation The conversion of digital signals to analog signals on a sending communications device.

MPEG-1 audio-layer-3 (mp3) Audio compression format that preserves original uncompressed sound quality when played.

multifrequency (MF) tones Combination frequency tones used for in-band signaling on the voice telephone network.

multiple subscriber line carrier system Commonly referred to as a *Slick-96* (SLC-96), this pedestal remote terminal device takes 96, 64-kilobits-per-second analog voice or modem signals, converts them to digital, and then multiplexes them at the remote terminal. The remote terminal is then connected to a central office terminal (COT) using five T-1 (DS-1) lines.

multiplexer Device used to combine multiple signals on a telecommunications network.

multiport repeater Ethernet network devices used to centralize network connections; at the repeater, a signal received on one port is repeated to all attached ports.

multistation access unit (MAU or MSAU) Token Ring collapsed-ring hub.

near-end crosstalk (NEXT) Electromagnetic signal crossover that occurs between a transmitted signal and a received signal on a copper wire pair. Transmitted signals are typically stronger than signals that are being received and interfere with the received signals.

neighborhood node Cable network connection point that can serve up to 1000 customers.

network address translation (NAT) Translation of an Internet protocol (IP) address used within a network to a different IP address used on another network using a firewall.

network administrator Person who has total control of a server-based network, controlling the access of all users with each user set up with a unique username and password.

network interface (NI) Connection box installed by a telephone company technician that is considered the point of demarcation between the installing telephone company and the customer.

network interface card (NIC) Device used to physically connect a computer to the network transmission medium.

network layer Layer 3 of the OSI model that provides the method for addressing and routing data when two different networks are connected together.

no acknowledgment (NAK) A reply back from a receiving terminal to the sender indicating that a piece of information was not successfully received.

node A communications network connection point device that is configured to process communications traffic.

noise In telecommunications, any part of a signal that is not desired.

non-facility-associated signaling (NFAS) Signaling technique that allows multiple ISDN PRI lines to be combined and controlled with one 64-kilobits-per-second ISDN D-channel.

NSFNet A network of scientific and academic computers funded by the National Science Foundation in 1980.

Nyquist sampling theorem Telecommunications theorem that states that an analog signal must be sampled at a rate of at least twice its highest frequency for sample reproduction.

open system interconnect (OSI) model Network structure for standard interfaces and communications protocols developed and promoted by the International Standards Organization (ISO: http://www.iso.ch).

open wire Uninsulated thick copper wire strung from telephone pole to telephone pole used to build the early voice network.

operator Human with the job of switching telephone calls from one point to another.

optical carrier (OC) SONET bit rate indicator. For example, the SONET standard bit rate is 51.84 megabits per second and is referred to as optical carrier (OC) -1 or synchronous transport level (STS) -1.

out-of-band signaling Signaling technique that uses a completely separate network from the one that carries voice; as a result, signaling occurs at a much faster rate. The SS7 network uses out-of-band signaling.

out-pulsing Local loop electrical pulse generation using a rotary telephone. Pulses are generated by opening and closing an electrical switch when a dial is rotated and released.

outside wire Telephone-company-installed telephone wiring that makes the local loop connection between a telephone being used in a residence or business and the local-exchange carrier (LEC) central office (CO).

overhead In a transmission system, anything that is not actual information (voice, data, video, etc.) that is transmitted along with the information.

packet assembler/disassembler (PAD) Used with frame relay, a computer-running software that will take data and break it down into packets for transmission.

packet-switched network A network designed to separate information into packets and not use dedicated circuit line connections like the PSTN.

packet-switching exchange (PSE) With reference to frame relay, a computer, also commonly referred to as a *node,* that routes the frames through the network cloud toward the destination.

parity checking Technique used for data-transmission error checking.

party line Early telephone connection that was shared between neighbors.

passband The range of low frequencies passed by an electronic filter with the limit determined by the cutoff frequency.

payload With reference to ATM, the 48 bytes of data within an ATM cell.

peer-to-peer Simple network configuration that allows individual users to administer and control network access to their machines.

permanent virtual circuit (PVC) With reference to frame relay networks, the connections formed within the frame relay network cloud.

phase-shift keying (PSK) Modulation technique that varies the phase of a carrier signal.

phreak Original telephone network hacker term used for people who learned to manipulate the voice network to obtain free long-distance calls.

physical layer Layer 1 of the OSI model where bits are put on and taken off the specified transmission media.

plain old telephone system (POTS) The existing telephone network that has been tuned to function at human voice frequencies; also referred to as the *public switched telephone network (PSTN).*

point of demarcation   Location on the telephone network connection that sits between the installing telephone company and the customer.

point of presence (POP)   Internet service provider (ISP) customer connection point.

preferred interexchange carrier (PIC)   Customer-selected preferred long-distance call carrier. The PIC is added to the local switch database the customer is connected to for connection and billing purposes.

presentation layer   Layer 6 of the OSI model that controls the conversion of information from the application layer.

primary-rate interface (PRI)   The integrated services digital network (ISDN) service that offers 23 digital 64-kilobits-per-second B-channels and one digital 64-kilobits-per-second D-channel.

private-branch exchange (PBX)   Device that switches telephone calls internally within a business.

proxy server   Device that is typically combined with a firewall and increases Web surfing performance by caching, or storing, visited Internet Web sites.

public switched telephone network (PSTN)   The existing telephone network that has been tuned to function at human voice frequencies; also referred to as the *plain old telephone system (POTS)*.

pulse amplitude modulation (PAM)   Modulated signal formed using pulse code modulation (PCM).

pulse code modulation (PCM)   Modulation technique that uses sampling to obtain instantaneous voltage values at specific times in the analog signal cycle to generate a pulse amplitude modulation (PAM) signal.

quadrature amplitude modulation (QAM)   Modulation technique that combines amplitude-shift keying (ASK) and phase-shift keying (PSK) on a carrier signal.

quadrature amplitude modulation 64 (QAM 64)   QAM modulation technique that provides downstream bandwidth of up to 36 megabits per second with eight bits per baud transmitted.

quality factor (Q)   The ratio of inductor reactive power to inductor resistive power in a resonant circuit.

quality of service (QoS)   An indication of a system's ability to provide a reliable service.

quantization   Telecommunications technique used—along with a sampling rate—to generate a pulse code modulated (PCM) wave.

quantizing noise   Signal distortion on the public switched telephone network (PSTN) caused by quantization, a voltage-level-adjustment technique that matches one of 256 discrete voltage levels.

rate-adaptive DSL (RADSL)   Digital subscriber line service that adapts its data rate to the level of noise and interference on a given line. Bit rates of 1.5 to 8 megabits per second downstream and 1.544 megabits per second upstream are possible over local loop distances of up to 1.5 kilometers depending on bandwidth.

read-only access Network sharing option that only allows others attached to the network to read files.

receiver Speaker device in telephone handset.

reed switch Electronic switch made of remendur and used to make the physical contact required to complete and disconnect a telephone call within an electronic switch.

Regional Bell Operating Company (RBOC) Term used to refer to the regional U.S. telephone companies (e.g., Verizon, SBC) that were created with the breakup of the American Telephone and Telegraph Company (AT&T).

reject region Range of frequencies not passed by a filter.

relay coil Device used in a telephone company switch to determine that a customer wishes to make a call.

remendur Magnetic alloy used in modern telecommunications reed switches. The alloy allows the reed switches to operate in two states, on or off, and the alloy is made magnetic by applying electric current through the coil in one direction and nonmagnetic by applying electric current through the coil in the opposite direction.

remote terminal (RT) Portable switching facility outside of central office, usually placed on a cement platform.

repeater Device that takes any incoming electrical signal on the repeater input line, amplifies it, and sends it back out on the output line.

request for comments (RFC) A document that includes definitions of Internet policy and protocol. All Internet communications protocols are based on RFCs.

resonant frequency The frequency, in a series RLC circuit, where circuit impedance is equal to only the resistance value.

ring Term used along with *tip* to designate the local loop copper pair wires.

ring network A network constructed in a ring topology with backbone ends joined together.

rotary dial service Electronic pulse-generation technique that uses an electrical switch to generate pulses on the local loop when a rotary telephone dial is rotated and released.

router Device used to separate logical networks and control routable protocol traffic.

satellite A combination transmitter and receiver that is launched into space and set in orbit around the earth and uses earth station ground dishes and antennas to communicate.

server-based network Network that has been designed for larger groups of users than a peer-to-peer network.

session layer Layer 5 of the OSI model that is responsible for establishing and maintaining communications sessions.

Shannon's law Telecommunications law that states that the maximum data bit rate of a communications system is determined by the bandwidth and the signal-to-noise ratio.

shielded-twisted-pair (STP) cable Twisted-pair copper wire shielded with foil or metallic braid to protect from electromagnetic interference.

Siamese cable Telecommunications cable commonly used by cable network companies that includes both coaxial and fiber cable. The coaxial cable is initially used with the fiber going unused. By running Siamese cable, the cable companies can easily extend the neighborhood node network out to new subdivisions and homes as they are built without running new fiber cable.

sidetone The sound of the telephone user's own voice heard in the telephone receiver allowing talkers to more easily adjust their voice volume when speaking into the telephone transmitter.

signal control point (SCP) Signaling System 7 (SS7) switch databases that work with the Signal Transfer Points (STPs). The SCP database provides information about advanced call-processing capabilities that are provided and used by an SS7 network.

signal switching point (SSP) Signaling System 7 (SS7) enabled switches at both the caller and receiver ends of a telephone call. These switches originate, terminate, or switch telephone calls.

signal-to-noise ratio (SNR) Ratio of a telecommunications system signal in decibels (dB) to the telecommunications system noise in decibels (dB).

signal transfer point (STP) Signaling System 7 (SS7) enabled switches that handle the long-distance path of SS7 signal traffic. These switches receive and route incoming signal messages to the required destination.

Signaling System 7 (SS7) Out-of-band telephone network used only for signaling. The SS7 network works in conjunction with but separate from the voice network.

single-carrier modulation quadrature phase-shift keying (SCM QPSK) Also referred to as *distributed queue dual bus (DQDB)*, a cable modulation technique that generates one of four possible constant amplitude states, with each separated by 90 degrees, and provides downstream bandwidth of up to 10 megabits per second.

single frequency (SF) tones Tones made up of only one frequency that are used for in-band signaling on the voice telephone network.

single-mode fiber Fiber-optical transmission medium with potential bandwidths of greater than 100 gigahertz. This fiber is used almost exclusively for long-haul telecommunications.

software driver Program that allows an operating system to access an attached a supplemental hardware device.

splitter   Filtering device used in ADSL systems to separate voice and data frequencies.

splitterless DSL   Digital subscriber line (DSL) service that runs at 1.5 megabits per second downstream and up to 512 kilobits per second upstream. Also known as *G.lite* or *universal DSL*.

stacking   The process of connecting Ethernet multiport repeaters to increase the number of Ethernet devices in a single collision domain; also referred to as *uplinking*.

star topology   network physical-connection technology that uses individual cable attached to a centralized connection hub for each device attached to the network.

station wire   Common residential telephone wiring that has four copper wires that combine to form two wire pairs. Also referred to as *inside wire, POTS wire,* and *JK wire*.

statistical time division multiplexing (STDM)   Multiplexing technique that allows multiple devices to communicate by adding an address field to each time slot in the frame. Empty frames are not transmitted; only devices that require time slots get them.

step-by-step switch   An array of Strowger switch frames that can handle 10,000 individual telephone lines.

step-index fiber   Fiber-optical transmission medium with relatively low bandwidth and high loss when compared with other fiber types; suffers from modal dispersion and not typically used in telecommunications systems.

Strowger switch   The first automatic mechanical switch—developed in 1892—that went into wide use in 1919 when the Bell System introduced dial telephones.

subscriber line carrier (SLC)-96   Device used to combine and multiplex 96, 64-kilobits-per-second analog voice lines.

subscriber loop   The connection between a telephone being used in a residence or in a business and the local-exchange carrier (LEC) central office (CO); also referred to as the *local loop*.

superframe   Grouping of 12 frames in the DS carrier system; also referred to as a *masterframe*.

supergroup   A frequency division multiplexed combination of 60 individual voice channels.

super-unit binder   Telecommunications cable that holds more than 24 binders (600 pairs).

switch   With reference to Ethernet, an enhanced network bridge that has been developed to reduce the performance bottlenecks associated with multiport repeaters.

switchboard   Telecommunications connection device, run by human operators, used to build telephone call circuits.

switchhook   Switch on a telephone handset that closes when a receiver is picked up, causing current to flow from the central office switch on the local loop.

symmetric digital subscriber line (SDSL)   Digital subscriber line service that offers upstream and downstream channels of 768 kilobits per second on a single wire pair on local loop distances of up to 4 kilometers.

synchronous optical network (SONET)   Transmission technique that transmits at rates between 51.84 megabits per second and 9.95328 gigabits per second.

synchronous transport signal level (STS)   SONET bit rate indicator. For example, the SONET standard bit rate is 51.84 megabits per second and is referred to as optical carrier (OC)-1 or synchronous transport level (STS)-1.

tandem office   Telecommunications facility developed to concentrate trunk switching and serve an area approximately the size of a small town.

terminating resistance   Resistors that are attached to each end of bus network cables to prevent transmitted signals from bouncing off the ends of the cables back onto the network.

termination   Used on Ethernet networks to prevent signals from bouncing off the ends of cables and interfering with other transmitted signals.

time division multiplexing (TDM)   Multiplexing technique that allows multiple devices to communicate over the same circuit by assigning time slots for each device on the line.

tip   Term used along with *ring* to designate the local loop copper pair wires.

Token Ring   An IBM product, Token Ring is the most common collapsed-ring network topology in use today; it uses either shielded-twisted-pair copper wire or UTP data cable.

toll center   An interconnect point for tandem offices set up to provide long-distance calling to customers; also referred to as a *toll office*.

toll office   An interconnect point for tandem offices set up to provide long-distance calling to customers; also referred to as a *toll center*.

toll-connecting trunk   Toll center interconnection transmission media.

traffic   Generic telecommunications term used to indicate voice, data, and/or multimedia transmission.

traffic density   The maximum traffic value for a defined *busy hour;* expressed in century call second (CCS) units.

transmitter   Device built into the handset of a phone that is responsible for converting sound waves into electrical signals.

transport layer   Layer 4 of the OSI model that provides reliable end-to-end communications and is responsible for ensuring that a communications session is established and then torn down, or removed, when the session is complete.

truck roll   Term used to indicate a telecommunications technician visit to a customer site.

trunks   High-bandwidth transmission links that carry telecommunications signals over long distances.

universal ADSL (UADSL)   Digital subscriber line (DSL) service that runs at 1.5 megabits per second downstream and up to 512 kilobits per second upstream; also known as *G.lite* or *splitterless DSL*.

universal serial bus (USB)   Personal computer (PC) device port that supports a data bandwidth of 12 megabits per second.

unshielded-twisted-pair (UTP) crossover cable   A specially constructed UTP cable that allows Ethernet multiport repeaters to be connected together into a single collision domain.

unshielded-twisted-pair (UTP) wire   copper cabling commonly used for local-area-network (LAN) construction.

uplinking   The process of connecting Ethernet multiport repeaters to increase the number of Ethernet devices in a single collision domain; also referred to as *stacking*.

very high speed digital subscriber line (VDSL)   Digital subscriber line service that provides between 13 and 52 megabits per second downstream and between 1.6 and 2.3 megabits per second upstream over distances of up to 1.5 kilometers depending on bandwidth.

wavelength   The distance in meters of one sinusoidal light wave cycle.

wavelength division multiplexing (WDM)   Multiplexing technique that uses wavelength, or color, of light to combine signal channels onto a single piece of optical fiber.

Web client   An Internet browser-editor. Examples include Netscape Navigator and Microsoft Internet Explorer.

wide-area network (WAN)   A network used to connect remote LANs located large distances away from each other into an enterprise network.

winding resistance   The resistance all inductances have due to wire windings within the inductor.

World Wide Web (WWW)   All the available resources on the Internet that use the hypertext transfer protocol (HTTP).

Index